The Social Regulation of Competition and Aggression in Animals

THE SOCIAL REGULATION OF COMPETITION AND AGGRESSION IN ANIMALS

Martin Moynihan

SMITHSONIAN INSTITUTION PRESS
Washington and London

EDITOR: Tom Ireland
PRODUCTION EDITOR: Jack Kirshbaum
DESIGNER: Janice Wheeler

Library of Congress Catalog-in-Publication Data

Moynihan, M.
The social regulation of competition and aggression in animals / Martin Moynihan
p. cm.
Includes bibliographical references and index.
ISBN 1-56098-788-x (cloth : alk. paper)
1. Competition (Biology) 2. Aggressive behavior in animals.
3. Social behavior in animals. I. Title.
QH546.3.M69 1998
591.56—dc21 97-53122
CIP

British Library Cataloging-in-Publication data available

04 03 02 01 00 99 98 5 4 3 2 1

♾ The paper used in this publication meets the minimum requirements of the American National Standard for Permanence of Paper for Printed Library Materials Z39.48-1984.

For my colleagues and competitors

CONTENTS

PUBLISHER'S NOTE

During the review and preparation of this book, Dr. Moynihan became critically ill. He devoted himself to the project right up to his last moments, especially in finding better and clearer articulation of the manuscript's knottier philosophical points. Before and after his death, we received many offers of help from the author's friends, students, and colleagues (categories that we found were widely overlapping). We were happy to receive this help. In particular, we acknowledge very significant assistance from Michael Robinson, W. John Smith, Eugene Morton, Bridget Stutchbury, and Olga Linares.

EDITORS' PREFACE

Martin Moynihan died on 3 December 1996, just before completing the finishing touches on his book. He left us all a gift. His memorable style is a part of it—witty, succinct, and peppered with mild sarcasm to keep you on your toes. You are assumed to be witty and conversant, and Martin never tries to explain in too much detail, for that would assume you aren't knowledgeable and probably not very bright either. You would be a bore at a dinner party. No, you are the object of his attention and admiration when you read this gift.

This book should be read by everyone interested in the evolution of behavior because Martin has outlined a new way of viewing the subject; indeed, he has presented us with a new subject. The book is not about competition or aggression but about the *control* of them. Aggression and competition must be controlled before any complex social behavior, at all, can exist. But this presents a dilemma: how can natural selection bring about controls on aggression and competition when it should be perfecting talents to engage successfully in them? This is the same sort of dilemma as the evolution of sex and the cost of meiosis (e.g., Eberhard 1985). On the face of it, it should not exist. Yet controls on aggression do exist, and this book is the first to focus on their forms and evolution. It is an important book because competition and aggression feature so prominently in the "struggle for existence" and in sexual competition—they determine the winners and losers in Darwinian fitness.

To refocus from aggression and competition per se to their controls might necessitate a new vocabulary, new terms and definitions. Martin spares us by using old and familiar terms, like *habituation* and *gregariousness,* and terms from ethology (e.g., *redirection*) that are defined in a glossary. The book is understandable to the beginner as well as those experienced in the jargon and concepts of evolutionary biology. His great insight is evident in how he describes habituation, for example, in light of the function for which it truly evolved—to reduce aggression. Describing habituation and other well-known concepts in this new light is something, as far as we are aware, that has not been done before.

He provides us with a first attempt to organize the tactics and strategies of controls by placing them into "simple" and "complicated" categories. Well, simple in form, at any rate. He illuminates the important ecological stages that explain patterns of the control of aggression. Many of these he has personally experienced and written about. He reports many observations here for the first time, and the observations of others are enhanced within this new framework. These patterns are expressed in descriptions of how and why aggression varies geographically among Andean birds, for example, and in animals with very lethal weapons, such as the "spear-billed" kingfishers and rollers. Animals may have "tooth and claw" for reasons of eating and other nonaggressive pursuits. How on earth, and why, are these potential weapons controlled; how are "guns turned into social plowshares?" We appreciate Martin's fascination with interspecific associations of birds much better after reading this book, because we understand the mosaic he was building. Of course, his own thinking had matured beyond his original papers depicting the communication and habitats among these animals (e.g., "social mimicry," Moynihan 1968). Martin needed to write this book to codify his experiences, his lifetime of *empirical* thinking and theorizing, the tradition epitomized by Darwin.

Martin presents his observations with great enthusiasm yet with proper restraint, homage to the complexity of behavioral evolution. Consider the talent of being a first-rate empirical theorist, that is, using observation and experiment to produce understanding. You must know the natural history of your subject and be conceptually and constantly aware of the logic of natural selection. With this as a start, you fit together your observations to determine how your living subjects were influenced by their ancestors and are now balancing constraints and contradictory sources of selection. (We envy the simple life of physiologists like T. H. Morgan, who chastised biologists for worrying about sexual dimorphism beyond its hormonal basis, content with knowing the mechanics of *how* something works, not *why* it exists in the first place [Mayr 1982, 72–73].) To be a first rate empirical theorist takes a lifetime of devotion and a great amount of experience with fieldwork and learning how to observe and be observant.

Pity the biomathematical theorist who only considers the cacophony of life from a desktop. The real world is more complex than can be imagined or modeled. With mild sarcasm, Martin also hints that there is something very wrong, flawed, in biological thinking these days, when bright young biologists position empirical work just slightly above natural history. They may be bright, but they don't know very much. They will not find this book enlightening because they don't appreciate empirical research, which means, automatically, that they don't care a whit about descriptions of animal behavior.

The "modern" approach to behavioral research is often embedded in a hypothesis-testing framework, with innovative ideas going unnoticed unless first rigorously tested with experimental manipulations. Thus, behavioral ecologists may initially be frustrated and impatient with Martin's detailed behavioral descriptions. But as he reminds us in his opening paragraphs, the details *are* what matters! Recently, behavioral ecologists have begun to realize their science lacks a foundation of the mechanics of behavior—how it works, its pragmatic effects. Two of five symposium speakers at a recent international meeting called for more interest in the mechanisms of behavior to augment a strong foundation in evolutionary theory. They suggested that mechanisms responsible for evolutionary change are going to be a hot topic of research in the near future.

The Social Regulation of Competition and Aggression in Animals brings us a whole new way of viewing social behavior. The regulation of aggression in the face of competition must be described, the actual behavior must be observed, before good predictions can be made and tested. Martin lamented the scarcity of behavioral details in most recent research. His book should encourage the continued resurgence of interest in and enrich our understanding of behavior. He provides needed tools. It is up to us to continue producing empirical theory while there is still something left of the natural world. In the end, all that will be left is desktop theorizing.

Eugene S. Morton
Bridget J. M. Stutchbury

FOREWORD

From 1957 until 1973, Martin Moynihan was resident naturalist and director, first of the Canal Zone Biological Area (CZBA), and then of the Smithsonian Tropical Research Institute (STRI). CZBA was comprised of the Barro Colorado Island field station and a one-room office in Balboa, Canal Zone. The CZBA was renamed STRI in 1966, the year, coincidentally, that I joined its staff. Under Martin's leadership STRI became a world-class research institute with a varied, enthusiastic, and peripatetic, tropics-trotting staff; two, and then three, marine laboratories; a spectacular tropical library; a research vessel; and a swelling number of talented visiting researchers. Barro Colorado attracted an international cast of characters, many of whom were *characters.*

Martin effected a punctuated, expansionist phase of evolution, perhaps a saltation, by sheer force of will, strength of character, and glory of vision. He was supported by S. Dillon Ripley, then the Smithsonian secretary, whose vision (almost) matched Moynihan's. The two men defined not only the size and ambiance of STRI, but also its scientific style. Moynihan's background was in ornithology. His first paper, coauthored with Ernst Mayr at the age of eighteen (he must have done the museum research at seventeen), was on New Guinea birds. After Princeton he went to Oxford, where he was infected by European ethology. The maestro, Niko Tinbergen, introduced Martin to *Larus ridibundus ridibundus,* the blackheaded gull—surely a day when the gods laughed. It was the start of a massive odyssey into the ways of gulls that culminated in *Revision of the Family Laridae* and more than a dozen significant papers. It is almost certain that his lifelong interest in the origins, functions, description, and comparison of displays started there. Gulls certainly are masters of "histrionic" displays, destined to make a lasting impression on gull and student alike.

The Oxford Animal Behaviour Group was then replete with the early ferment of "missionary zeal." A whole generation of Tinbergen disciples coruscated with ideas—including Desmond Morris and Moynihan, David Blest and Aubrey Manning. Within range of Oxford's dreaming spires, the Cambridge group, led by Robert Hinde, provided a competitive stimulus. These were the frontier days of English-speaking ethology, and Martin brought back with him a augmented enthusiasm that never waned. Few people steer toward a large institution *and* continue research at a high level of excellence. Martin did. In his seventeen years as director, he published twenty-five papers—not mere notes—over one thousand pages in total. Among these were massive treatments of his gull research, his early papers on Andean birds, and his book on New World primates. By the time he resigned as director, he had started his fieldwork on cephalopod behavior, the first summary of which appeared in 1975. Let me emphasize the amazing reality of this achievement: While building STRI he launched, and carried out, his important studies of Andean birds, New World primates, and cephalopods. In addition, he published a series of major theoretical papers, including the one that I think is "first among equals" in its implications for Evolution, *The Control, Suppression, Decay, Disappearance and Replacement of Displays* (Moynihan 1970b). His studies of the reef squid, *Sepiateuthis sepioidea,* proved to be marvelously productive. Here was an animal with an extraordinary capacity for complex visual displays of pattern, texture, and color studied by an extraordinarily perceptive master of display studies. The skin of the squid could flash patterns on and off as rapidly as a cathode ray tube, the tentacles and arms could assume a wide range of postures, and the skin could change from smooth to papillated. On top of all this, Martin compared displays of the three major orders within the cephalopod subclass Coleoidea. Because these orders are known from fossil evidence to have separated in the Mesozoic, this evidence provided a unique marker for behavioral evolution. He concluded that some of the displays were *at least* 100 million years old. "Extremely conservative," he would say, with a tendency to understatement perhaps acquired at Oxford. This work seems to have been passed by, relatively unnoticed, in the flurry of now more fashionable behavior studies.

When he became senior scientist in 1973, he continued to study displays and their significance. His marriage to Olga F. Linares, a very distinguished anthropologist, resulted in both a contentment and a synergy that further enhanced his insights and productivity. They eventually bought a country house in Albi, France, where between sojourns at STRI in Panama and research trips to Africa to study kingfishers and their relatives, Martin studied pheasant behavior, publishing a paper on Reeves's pheasant in 1995.

Martin Moynihan lived and worked largely outside his home country and its population of ethologists. I remember making a trip with him to the Amazonian forests of Colombia, he to study primates, I to search for cryptic and mimetic insects. Our field styles proved to be totally inimical. Studying monkeys means rushing through the forest at breakneck speed to encounter troops of monkeys and watch them until they move into inaccessibility, and then repeating the process. I, of course, needed to go slowly and search under every leaf. We saw sakis and titis, magnificent in their natural setting. I tried to keep pace until heat prostration hit, when Martin showed his deep sensitivity, finding a nearby hut where I could be rehydrated and covered with cooling wet towels. How easy it would have been to be impatient with an overweight neophyte.

Later, I remember watching my captive predators, rufous-naped tamarins, with Martin at my shoulder. I'd been working with them for several months, running them several times a day through experiments on prey detection. I thought I was one hell of an observer as a result of Niko Tinbergen's tutoring; but I hadn't the Moynihan eye. Martin watched a display of female tamarins that I had interpreted as a copulation rejection display. The tamarin flung out her tail and conspicuously rolled it up into a catherine wheel, pressed tightly against the external genitalia. This seemed "designed" to block male access and say no. But after watching it several times, Martin decided that it had just the opposite function—invitation, not turnoff. He was right. If I'd studied the male response, that would have been obvious. Clearly, as he would say, I had a lot to learn, and I hope I did.

Snorkeling next to Martin during his squid work in the warm and pellucid waters of Panama's San Blas Islands was a joy. As an underwater observer, he was a bizarre sight. He wore snorkel and mask, tennis shoes, and a belt accoutered with underwater note pads, writing and drawing implements, and an old tobacco pipe: the Christmas-tree effect. I don't know whether between observation bouts he blew the salt water out of the pipe and smoked, but I can't think what else it was for.

An important pedagogic ritual was the Moynihan dinner party, alluded to in the editors' preface. Scientific guests were always assembled in a mix that guaranteed intellectual jousting of a high order. The ambiance ensured good food and the loss of inhibitions. For each of us, the *veritas* was assured by the *vino*. What finally resulted was a free-wheeling discussion that must have come close to recreating the best points of a tutorial at a medieval university. Ideas were aired, born, expanded, opposed, conflated, debated, and often destroyed, but everyone was the better for it. It certainly did more for critical faculties than the best seminars I've ever attended. In the end, after midnight, we always got on to a taxonomy (or perhaps the systematic) of biologists—"present company excepted." The species were first class, second class, and so on, with the subspecies first class, etc. "First class first class" was the highest accolade. Usually, only Darwin occupied that category, but sometimes Niko and/or Konrad were admitted to the pantheon. This had to be a delight for all who took part—hundreds over the years.

The significance of this book is assessed by Morton and Stutchbury in their preface. Their views are lucid and apposite, and they correctly state the importance of this closely argued new look at displays. I should not add to their comments, but the temptation is overwhelming because I suspect that Martin's approach is not part of the mainstream emphasis, nor does it reflect the conventions of a generation or so of young ethologists. Far-reaching comparisons such as these are unfashionable, but if they are read as a source of so many interesting, fundamentally important hypotheses, they will certainly stimulate research. In a very real way, Martin has done what Darwin did in *The Origin of Species.* He has accumulated a mass of information and drawn major and radical conclusions from it. And all these ideas are testable, and eminently worth testing. Providing data to fill the gaps to which Martin draws clear attention, over and over again, is a daunting prospect. On the other hand, the results would surely advance our understanding of the evolution of behavior. There could be no better memorial for any person.

Shakespeare gives me an appropriate exit line, summarizing all the above: "He was a man. . . . I shall not look on his like again."

Michael H. Robinson
Director, National Zoological Park

INTRODUCTION

Both competition and aggression are facts of life.

Competition would appear to be ubiquitous, universal, inevitable. All organisms compete with others. Categories of "others" include family, friends, neighbors, acquaintances, rivals, and enemies. Of course, not all the categories are always mutually exclusive. The objects of competition are exceedingly diverse. They include anything and everything that might contribute to personal or to inclusive fitness: the reproductive success of individuals, or the survival of their genes in their own offsprings or those of their relatives.

Aggression is a character of animals. It may—perhaps must—occur in all animals of any appreciable complexity living in anything like natural conditions.

Competition and aggression often occur together in time and space. This is the principal reason why they can and should be discussed together. The association between the two phenomena is so frequent, and often so close, that we may reasonably suppose that they are related to one another causally. But why and how? The details are important.

One should ignore Voltaire in this context. Details may or may not kill great works. They certainly do, however, decide the fates of organisms.

The necessity for competition is evident. Resources are limited as well as limiting. The necessity for aggression is perhaps less obvious at first glance. Yet it is quite real. To put the matter as simply as possible, aggression seems to be *primarily an adaptation to cope with competition*. It seems, in fact, to be the principal *social* control of competition. It can discourage, repel, or suppress enemies and rivals. It may be a first recourse. In any case, it is usually the strongest and/or ultimate active recourse. (It is not, of course, the only control. Some examples of nonaggressive controls of competition are noted below.)

Although the relations between competition and aggression are simple enough in theory, they can be complicated in fact. One complication can be mentioned briefly. There are different kinds of hostile encounters. Two main divisions are likely. Interference competition usually involves overt, physical confrontation. Exploitation competition may be mediated by distant signals, such as conspicuous displays and songs; but these patterns themselves are often hostile in motivation. Forms of aggression may be more subtle in the latter category.

All characteristics probably have both advantages and disadvantages. They probably are, therefore, exposed to different, often opposing, selection pressures. The pressures usually are unequal. Compromises can be found in some cases (on the one hand this, on the other hand that), but not in all. It is the precise nature and limits of the pressures, perhaps transitional, not necessarily stable, that need to be determined.

Some plausible assumptions may be taken on faith unless or until disproved. Throughout this book, it is presupposed that existing behavior patterns are or were "adaptive," that is, useful, on the whole more positive than negative. Their advantages must be, or have been, greater or stronger than their disadvantages. They were selected for because they resulted in relatively better survival and/or reproductive success. This assumption has been questioned by some scientists, for example, Gould and Lewontin (1979) and Cracraft (1990). There have been discussions of the subject. The pro-adaptationists, for example, G. C. Williams (1966), Frazetta (1975), Mayr (1982, 1983), Ridley (1983), and Queller (1995), would seem to have won the argument (also see Cronin 1991; Crawford 1993; Curio 1994; Dawkins 1996; and chapter 3 below).

Both competition and aggression are expensive of time and energy. Aggression has another possible drawback. It can be dangerous. Attacking or fighting individuals run the risks of injuries, wounds, or even death. Injuries are common. Immediate death is rare, but it does occur. One can even imagine circumstances in which a willingness to fight to the death would be encouraged and accelerated (Enquist and Leimar 1990). Everything would depend upon the relative values of the resources being contested for and the value of the future for the contestant. The prize would have to be glittering, and/or the prospects of the contestant would have to be really unpromising. An individual with an otherwise very dim future might be well advised to go for broke.

Be that as it may, most opponents should try to minimize risks in most circumstances. Even minor injuries can result in significant losses of inclusive fitness.

What can be done about risks? What are the options?

The one thing that seems to be impossible to do is to suppress or eliminate all expressions of aggression entirely. The tendency to attack can, however, be reduced, redirected, modified, controlled. Some palliative devices have been known for a long time to both animals and human observers.

Our knowledge of social relations is still incomplete, but information is accumulating rapidly. Enough is already known to be interesting. Perhaps we can begin to identify some correlations and to draw up some general rules. As a consequence, we might be able to identify, at least by implication, some subjects and problems that would well repay further investigation.

COMPASS OR SCOPE

The scope of the book should be defined. We can begin with a disclaimer. It is important to specify what the book is *not* about. It is not a general account of either competition or aggression as such. The aim is elsewhere, on a slightly divergent path. The role of aggression as a control has been noted. So has the need for controls of aggression in turn. The necessary checks developed during evolution. It is these checks that are the real focus of the descriptions and discussions in the following pages. It would, I think, be fair to say that most of the book is concerned with *the second-order controls, palliatives, of a first-order control, aggression itself.* Special attention will be paid to immediate or proximate mechanisms.

BACKGROUND AND DEFINITIONS

Books and papers on competition have been so numerous that they can hardly be cited individually. Lists of references are given in the works of Colman

(1982), Rubinstein and Wrangham (1986), Huntingford and Turner (1987), and Slobodchicoff (1988) (also see Dawkins 1976, 1982; Connell 1983; and Trivers 1985). For our purposes here, competition among animals may be said to occur whenever one individual occupies or preoccupies a resource that would otherwise be available to, and possibly or probably appropriated by, another individual of the same or another species. Lazarus (1982) broadly and sensibly defines a resource as something in the environment that is *potentially utilizable* by an animal (my emphasis). In actual fact, the resources include many kinds of opportunities and accesses to food, air, water, reproductive partners, breeding sites, social companions, display sites and periods, observation posts, hiding places or refuges, secure escape routes, and so on, almost ad infinitum. The most significant resources are different for different individuals of the same or different species in different circumstances.

Two competitive processes, previously mentioned, have been called "interference competition" and "exploitation competition" (Maurer 1984; Merila and Wiggins 1995; Minot 1981; Minot and Perrins 1986). The first is more or less direct. It may include face-to-face encounters between competitors. The second, exploitative competition, is more or less indirect. The resources in play or at risk are disputed by long-term or long-distance maneuvers. The two processes are different, but not always distinct. Both include aggressive components. There are many intergrading and intermediate performances.

Interference competition can be conspicuous, but it seems to be less widespread than exploitation competition. Certainly, it is the latter (with some of the intermediate forms) that has most engaged the attention of ecologists (e.g., Roughgarden 1983; Miller 1968; Case and Gilpin 1977; Schoener 1982; Carothers and Jacsic 1984).

Competition, in general, can hardly fail to be "limiting." Sooner or later, any group, population or species, will fail to increase because some parameters of the physical, biological, or social environments have been fully appropriated or exhausted by themselves or others. Malthus must be right in the end (also see Davies 1982 for a more behavioral discussion of the effects of competition on populations).

Organisms cannot avoid competition, and successful competitors are rewarded. But there is a price to be paid. For animals, the mere efforts of competition, the acts involved, are a cost, an expenditure of time and energy. An individual will, therefore, benefit from maximizing good effects and minimizing bad effects. It should strive to enhance and refine its own competitive abilities while, at the same time, weakening or controlling the competitive efforts of others. This is a matter of regulation. It can only be managed by or through behavior, action and reaction.

Aggression is another broad term. Like competition, it has been discussed in a number of general books over the years. There have been popular accounts, for example, Ardrey (1966). More seriously, one could cite, almost at random, J. P. Scott (1958), Lorenz (1963), Carthy and Ebling (1964), again Hungerford and Turner (1987), and Archer (1988). More recent studies have been narrower, if sometimes more profound, or largely concerned with human aggression (e.g., Karli 1991; and Silverberg and Gray 1992). The behavior of man, as such, will not be considered in this book.

Aggression may be said to include overt attack in all its forms, ranging from violent blows or strikes to simple "intention movements" (Daanje 1950) such as unfriendly advances toward a rival or opponent. To these should be added "mixed," ambivalent performances, in which elements of attack can be detected or inferred.

There are many kinds of mixed patterns. One kind, or series of related kinds, is particularly common. Stimuli that activate a tendency to attack usually release a tendency to escape as well. The results are hesitations and special signals such as threat and appeasement. As used here, *tendency* is probability of performance (the equivalent of *drive* in some of the older literature). The linkage of attack and escape is eminently practical. Still, the two tendencies often differ in strength. Either one can be obviously preponderant over the other. Again, all types of performances can be intraspecific and/or interspecific.

Aggressive patterns, with or without escape elements, are often called *agonistic* in the behavioral literature. This raises a stylistic or etymological point. Many terms used in the specialized literature differ slightly in meaning from the same words used in ordinary, everyday converse. In other cases, terms have been changed in transcription from a foreign language. *Agonistic* is a case. The Greek *agon* from which it is derived implies competition, even cooperative competition, rather than fighting. (Note its use in classical tragedy: *protagonist* and *agony*.) Here, I will generally use *hostile* instead of *agonistic,* but the two terms are virtually synonymous.

Most terms are used broadly here. One of the exceptions is *conflict,* which will occur often below. (The term *contest* may be preferable as a broader or more general label.) It is, of course, permissible to speak of "conflicts of interest" between males and females or between parents and young (Trivers 1974; also see Svensson 1995; and Bateson 1995). When the term is used here, however, it usually or often implies the performance or imminent probability of actual fighting.

DIFFERENT APPROACHES: ETHOLOGY COMPARED TO SOCIOBIOLOGY

Analysis of the material will be ethological—as near to classical ethology (Tinbergen 1963) as possible. Not very fashionable at the moment. Dewsbury (1992) discusses some current trends. Curio (1994) is of the opinion that recent ethology has become almost cryptic. This may be an exaggeration, but the classical approach certainly has been overshadowed by "sociobiology," a rather disparate combination of purported socioeconomics with intermittent exercises in "game theory." Game theory itself, a serious discipline, is explained in rigorous mathematical terms by Straffin (1993). As invoked by sociobiologists, it is not always so rigorous. Early publications in the field include Alexander (1974), Wilson (1975), Maynard Smith (1974, 1979, 1982a), and Barlow and Rowell (1984) (also see the collection in Barlow and Silverberg 1980). Later papers can be found in almost every recent issue of journals on behavior and behavioral ecology, as well as in the publications identified by their titles as strictly sociobiological.

The basic difference or *décalage* between classical ethology and typical sociobiology might be summarized in a small series of antithetical, dialectical phrases. Sociobiologists are interested in the ultimate or long-term whys, including goals and strategies. Ethologists may be interested in the same problems, but they are more immediately concerned with short-term, even day-to-day, hows—in other words, with tactics. Sociobiologists want to know why certain social arrangements have been selected for over generations. Ethologists usually assume that natural selection is impeccable (on logical grounds, it can hardly fail to work, if not always perfectly). But they do want to know how particular arrangements are started, maintained, or discarded, now, in the real world. Who is doing what and to whom? With what effects? (Also see Shuttleworth 1995.)

There is a famous game that has become almost a hallmark of sociobiology. It is usually called "the prisoner's dilemma." Essentially, a prisoner must decide to cooperate with or defect from (cheat) fellow prisoners. The decision is strategic, dependent upon circumstances, and an appropriate subject for sociobiology. An ethologist would like to know what the players are doing and saying while they spin out their plans and plots, and what effects their activities actually have (not *might* have) on the final results. (When I was young, the game was called the "Spanish" prisoner's dilemma—presumably a distant reminiscence of the Carlist Wars. The dropping of the original adjective may have been an early example of political correctness upon the collapse of the Franco regime.)

Sociobiologists stress the significance of evolution. They are quite right to do so. But they have tended to concentrate on some aspects, the end results, more than mechanisms of accomplishment. The balance should be redressed.

Strategies are important in evolution. They can only work through or by tactics. There must be close relations between the two phenomena. It will be one of the principal objectives of this study to trace some of these relations in more or less precise detail. In the end, it will be seen that relations are complementary but not always predictable a priori. They can be remarkable and even surprising.

LEVELS OF EXPLANATION: THE SEARCH FOR SIMPLICITY

The physical forms of many actions and reactions can be described in such terms as *simple* or *complicated* or *intricate.* These terms are general and useful. They certainly are used often enough here, in descriptions and chapter headings. Unfortunately, some of the same terms have been used in previous publications by other authors in arguments as well as in descriptions. Complications have been posited too frequently in analyses of causation. There have been all too many references to "subtle" connections. This may have been a mistake, or at least misleading. However subtle or devious the relations between tactics and strategies may appear to be, the basic elements of both are essentially simple. Individual behavioral elements of tactics and strategies should be identified as such.

This approach may appear to be reductionist. But there are no reasonable, honest, intellectual alternatives available. Simplicity has to be looked for. It may be accepted religiously. It should also be helpful and explanatory.

EXAMPLES

Many animals will be cited in these pages. An effort will be made to include both old and new references.

The emphasis will be on the "higher" terrestrial vertebrates. They have provided abundant materials for observation and analysis of many different kinds of behavior. Given my own research, I am most likely to cite birds and mammals. Given the published literature, the mammals cited most frequently are primates. (The taxonomy of primates has been confused by a proliferation of "splitting," but there is a sensible and fairly recent classification in Kavanagh 1983.) Some studies of other mammals, especially rodents and carnivores, are also pertinent and suggestive (see Ewer 1968). Among rodents, several burrowing types have been studied at considerable length in the field or in seminatural conditions. Among carnivores, the hyenas, a small but diverse group, and a highly specialized mongoose are particularly interesting, instructive, and amusing.

Birds and many mammals place very considerable reliance upon sight and sound. Roughly speaking, their sensory equipment is similar to our own. Their performances are, therefore, more or less easily accessible to us. (The phrasing is vague on purpose. There are real differences among species. Thus, for instance, some birds can see into the ultraviolet. And, of course, many other mammals can hear much higher frequencies of sound than we can. Still, I think that the generalization will hold.)

Most of the species of birds and mammals cited are endowed with potentially formidable weapons. They have to manage aggression carefully.

From time to time, in the course of this account, I will also refer to coleoid cephalopods: the squids, cuttlefishes, octopuses, and their relatives. These animals are fiercely predatory, with all the necessary equipment and weapons, which they can use at will. The species living in well-lighted inshore waters have remarkably elaborate and sophisticated repertories of visual signal patterns. (There are general accounts of coleoid behavior in such works as Packard 1988a, 1988b; and Moynihan 1985, 1996. Discussions of individual species include, among others, Packard and Sanders 1971; Hanlon 1982; Hanlon et al. 1994; Dimarco and Hanlon, in press; Moynihan and Rodaniche 1977; Rodaniche 1985; and Mather and Mather 1994.) Coleoids are ideal for studies of neurophysiology, memory acquisition, and some, but not all, aspects of communication. At the moment, they are less than ideal for studies of aggression per se. Only a few species can be followed in the wild. Others have had to be kept in the inevitably somewhat artificial conditions of the laboratory.

I am not competent to discuss the "lower" vertebrates. All I can say, after brief glances at a few particular reports and general surveys (e.g., Carpenter and Ferguson 1977; Gans and Tinkle 1977; Miller 1978; Rand and Ryan 1981; Burghardt and Rand 1982; Ryan 1985; Rand 1988; Pitcher 1993; Martins 1994), is that the general principles of social behavior seem to be much the same in all vertebrates, including fishes, amphibians, and reptiles, given the different habitus and sensory capacities of the different groups.

I am equally ignorant of the huge and multifarious literature on the behavior of arthropods, the crustaceans, insects, and arachnids. In all probability, however, the results of many studies of these animals, having become part of the general consciousness of ethologists and sociobiologists, will be reflected, indirectly and at several removes, in various passages of the following pages.

Among birds, special attention will be paid to some members of the nominal order Coraciiformes. The group is supposed to include the hoopoes, wood-hoopoes, hornbills, rollers, ground-rollers, cuckoo-rollers, bee-eaters, kingfishers, motmots, todies, and perhaps trogons (comments in Sibley and Ahlquist 1972, 1985, 1990). I have studied some of these animals in parts of Africa, Madagascar, southern Asia, New Guinea, and the South Pacific (Moynihan 1987a, 1987b, 1988, 1990; see appendix 1). African species have also been observed and analyzed by many other students (see, for instance, Brosset 1983; Douthwaite 1973, 1978; Emlen 1982a, 1982b; Fry 1980, 1984; Kemp 1976; Kilham 1956; Ligon 1983; Reyer 1980a, 1980b, 1984; Thiollay 1985). The Palearctic kingfisher is described by Boag (1982). Australian forms are covered by Forshaw (1983, 1985). There is an extensive, if still only partial, review of Old World species in Fry et al. (1992). Comparable observations of New World species are summarized in Davis (1985), Keppler (1977), Scott and Martin (1983), and Skutch (1954, 1957, 1967, 1983).

Why have coraciiform birds attracted interest? There are several reasons. Many species are abundant, ecologically significant, and easily observed. More important in this context, they have varied social patterns and structures.

Most of the living members of the order are tropical. The tropics are remarkable for their biological diversity, and the nature of the diversity has been another subject of argument. We do not know if there are more species, on land, at any single point in the tropics than at any single point in the richest habitats of the north or south "temperate" zones. It may be reasonable to assume, however, that there are more species at adjacent points in the tropics than elsewhere. It is possible, but by no means certain, that competition is somehow particularly intense or frequent in tropical areas. If this is indeed the case, then one might expect aggression to be highly developed in the same areas. This would, of course, be relevant to coraciiforms as well as other animals. But is it true? The available counts of the relevant interactions are not always very useful. They were collected by different people, operating on different materials with different techniques, at different times and in different places. Even if competition in general should not be particularly intense in the tropics, some critical resources might be in relatively short supply. There is some evidence for this.

The problems confronting coraciiforms are increased or exacerbated by two disparate, incommensurable but not unrelated, features of their biology. Many of them, for example, most rollers and kingfishers, are high-order predators. They feed on fairly large prey. They use their bills to catch large arthropods, caterpillars, millipedes, centipedes, beetles, and orthopterons (among the few types that I can recognize) and such vertebrates as small fishes, amphibians, and reptiles. Their bills are—presumably therefore—relatively large and powerful, perhaps too powerful to be used in real fighting without extreme provocation or except in sheer desperation. Furthermore, almost all coraciiforms nest in holes: in trees, termitaria, wasp nests, earthen banks, stretches of bare ground. The holes usually are made, excavated, rather than found. Nesting birds chisel, hack, pound, and dig with their bills. The substrates attacked tend to be hard or resistant. This must provide another selection pressure for strength and power and, therefore, caution.

Competition for nest holes is significant. It can be deadly for some small birds in the north temperate zone. Collared flycatchers, *Muscicapa ficedula,* will literally die in struggles for holes with stronger birds such as titmice, *Parus* spp. (Merila and Wiggins 1995). This is interference competition at its most extreme. For the flycatchers, holes are both indispensable and effectively in short supply: perhaps the ultimate in prizes.

The struggle is not always so tragic or dramatic, but it can still be serious. Short (1973, 1979) notes that the problem of finding suitable holes or sites for holes can be difficult for birds of the order Piciformes, the woodpeckers and their allies. The Coraciiformes also are hard pressed (Fry 1980; and pers. obs.). Not only do birds compete for holes among themselves, but they must also try to cope with mammals. In West Africa, in Sénégal and Liberia, small hornbills of the genus *Tockus* sometimes find that holes that they may have dug themselves have been invaded and (pre)occupied by true squirrels of the family Sciuridae, or "African flying squirrels," of the family Anomaluridae (again, pers. obs.). This is interference competition in a picturesque form.

The habit of nesting in holes can or should be compared with the habit of building "constructed" nests in the open or semiopen. The choice between the two alternatives would seem to have had far-reaching evolutionary consequences, not all of which are very closely relevant to competition and aggression as the terms are used here. Still, they have some bearing on the matter, and a brief excursus is attempted in appendix 2.

To return to a main argument . . . Many animals "need," that is, would or do find it advantageous, to manage aggression, both their own and that of others. They do so in different ways in different circumstances. Chapter 1 outlines the simpler forms of management, some of which have not been viewed as controls of aggression per se.

1
TACTICS: THE SIMPLE IN FORM

Behavioral controls of aggression are so numerous and diverse that they cannot be listed in any very neat or progressive order. Still, one has to begin somewhere. For practical reasons, it is convenient to begin with the comparatively simple in form, and then to move on to the less simple. Or perhaps from the nonpeculiar to the slightly peculiar to the very peculiar in appearance.

As indicated above, the use of these terms, as morphological reference points, does not necessarily imply anything about causation (or about functions, for that matter—see chapter 3).

HABITUATION

The phenomenon of habituation usually "looks" simple indeed. It is manifested by, or can be inferred from, decreasing responsiveness to repeated or continuous stimulation.

Actually, the same or similar results can be produced by two (or more) different processes. In some animals, there is sensory adaptation. A receptor, perhaps only a single neuron, becomes "fatigued" and progressively less sensitive (Prosser 1981). The change can be said to be peripheral. Some of the same animals also have mechanisms, perhaps filters, in their central nervous systems that can (also) reduce or stop responses. This is the process to which the term *habituation* is usually applied (Peeke and Petrinovich 1983).

Decline of responsiveness is not very exciting to watch. Not surprisingly, it has seldom been a major concern of ethologists and psychobiologists.

Some degree of habituation or similar processes must, however, play a part in the "dear enemy" phenomenon. The term seems to come from Fisher (1954). Territorial rivals, originally wildly hostile to one another, may fight less frequently or less vigorously as they come to know one another. Wilson (1975) cites a number of bird species that behave in this way, but the phenomenon certainly is widespread among mammals as well as birds (see, for instance, Ewer 1968 and all the references to primates below). Something similar also occurs in at least one fish (Miller 1978). Familiarity must be equally important in reinforcing the cohesion of all sorts of sustained groups, dominance hierarchies, family units, flocks, herds, troops, quite apart from spatial and territorial arrangements.

The initial stages of habituation, when contestants are not yet entirely sure of one another, must be difficult to navigate successfully. There may be shoals and rocks in the way. There may be upsets. Progress may be slow. Habituation cannot occur instantly. It takes time because repeated stimulus presentations are needed before responses might wane (see later notes on coraciiforms and Andean birds).

If and when it is achieved, habituation may lead to a decrease, even disappearance, of aggression. It may result in peaceful coexistence or even cooperation. Senar et al. (1990) observe that familiarity breeds tolerance (presumably with or without contempt). This is a "good thing" from one point of view. Habituation may appear to be risk free. On the other hand, it does allow or facilitate competition, and therefore virtually encourages it (e.g., more animals can occupy the same space). Tolerance has its price.

RETREAT

There are other ways to try to play safe, apart from, or in addition to, habituation. A cautious or reluctant contestant can be more or less pacific, even ingratiating, or openly cowardly.

As noted above, pacifism can only be partial. Some animals have moved further in this direction than others. Social relations with minimal hostility are perhaps most likely to be established when potential opponents are very different and unequal. Or, conversely, when they are as nearly alike as the peas in a pod. (As far as I know, serious disputes are comparatively infrequent in the enormous herds of nearly indistinguishable wildebeest, *Connochaetus taurinus,* on the plains of East Africa.)

Appeasement is another and more promising possibility. It is common only in certain types of encounters, above all, during "courtship." It takes various special forms. It will be discussed below in connection with displays and affiliation.

Cowardice is a third option. An animal can usually protect itself from aggression by retreating more or less promptly and rapidly from an opponent. This is a direct solution. It is easy to perform unilaterally. It does, however, have the disadvantage of removing the escaping individual from resources in the immediate vicinity of the opponent, presumably the very resources that are being competed for. Thus real escape has its own drawbacks.

There may be fewer disadvantages to carefully controlled retreats of limited extent, just moving back or slightly out of the way. This can be, in some cases, apparently unilateral. The carrion eaters of East Africa provide several examples. Spotted hyenas *(Crocuta crocuta)* retreat before lions *(Panthera leo);* jackals (*Canis* spp.) retreat before spotted hyenas; vultures (spp. of *Gyps, Pseudogyps, Necrosyrtes,* and *Neophron*) discourage or displace (replace) one another in predictable ways and sequences. The hierarchy is obvious. (See, for instance, Williams 1964; van Lawick and van Lawick-Goodall 1970; and Kruuk 1972 for the general picture; and Wallace and Temple 1987 for details of the vultures.)

SPATIAL AND TROPHIC OVERLAPS, AVOIDANCE, AND EXCLUSION

There also are bilateral and multilateral retreats. They can be repeated again and again after interruptions. Sequences can be added to one another without necessarily transforming the valence of any component pattern per se. (This is an oversimplification. Position effects are ignored for the moment.)

It is possible to recognize two kinds of repellent or repulsive behavior under the labels of *mutual avoidance* and *mutual exclusion.*

Exclusion is often fairly long term. Potential competitors can occupy different and nonoverlapping, if often adjacent, areas for long periods of time—months or years (perhaps repeatedly in the case of migrant birds).

The separate areas may be defended as territories. Defense is likely to include overt aggression, complemented by caution or restraint, a reluctance to cross boundaries, or even a willingness to retreat when necessary. These various processes probably cannot be managed easily or well until they become reciprocal, at which point, we are back to the "dear enemy." The mechanics of exclusion appear to be almost entirely behavioral. Aspects of ecology are more often ultimate than proximate. One point is clear. Any individual that excludes others is not sharing.

Mutual avoidance can be long and/or short term. Actual or potential competitors may occupy the same or broadly overlapping territories more or less continuously. Within these areas, they do share many of the same resources. They do so carefully. Although there are exceptions such as ravens, *Corvus corax* (references in Heinrich 1988), sharing is not usually absolutely simultaneous. Still, in most cases, the division of spoils is frequently recurrent, sometimes at short intervals. An individual takes all or part of a resource at a given site. Then it leaves. Perhaps within seconds or minutes, another individual comes to the same site and tries to use the same resource. It may or may not be successful. Eventually it leaves in its turn. Then a third individual appears—or the first individual returns. These successions may occur again and again. As in exclusion, factors of aggression, restraint, and a willingness to retreat seem to be involved. Only the mix and localization of the patterns are distinctive.

Exclusion is common during intraspecific competition. Avoidance is common during interspecific competition. The two sets of tactics are not, however, completely restricted in occurrence. At times, similar responses appear in different situations, and different responses in similar situations. There may even be some geographic variation. The examples that follow include very different types of birds in different parts of the Old and New Worlds. They illustrate how the dynamics of exclusion and avoidance may overlap in space and/or time.

Some Andean Birds

One distinctive series of species is described in Moynihan (1979). It is the "Diglossa cluster" (several species of small, nectarivorous birds that flock together) of the high-altitude humid forest and scrub zone of the Andes stretching from western Venezuela through Colombia, Ecuador, and Peru to northern Bolivia. Aspects of the biology of the region have been reviewed by contributors to Veuilleumier and Monasterio (1986).

A general comment may be useful. The relevant areas, habitats, and biotas are part of the "cold tropics." Somewhat ridiculously, the tropics are distinguished from the so-called "temperate" regions of the world by being more equable in temperature. From a biological point of view, it does not always matter if the temperatures are high or low so long as they are fairly consistent throughout. There are parrots, flamingoes, tinamous, large carrion feeders, and other representatives of supposedly tropical groups in the freezing but equable cold of Patagonia and Tierra del Fuego. Consider also the high mountains of Central America, especially Costa Rica and western Panama (Skutch 1967, 1983).

The particular Andean cluster cited here (another will be cited later) includes flower-piercers (species of the genus *Diglossa* itself); the phylogenetically closely related conebills (*Conirostrum* spp.); both members of the order Passeriformes, suborder Oscines; several less closely related species of hummingbirds, family Trochilidae; supposedly the order Apodiformes; plus a miscellaneous assortment of other birds. The critical forms are nectarivorous-cum-insectivorous. As usual among most nectarivorous birds (see, for instance, Ford and Paton 1985), relations among individuals are actively and conspicuously hostile.

Some general discussions, for example, papers in Hinde and Groebel (1991), attempt to distinguish between "prosocial" and "antisocial" behavior. The distinction may not be always useful. "Pro" and "anti," friendly and unfriendly, are often associated, even combined or mixed. It could be argued that all interactions among individuals (apart from predator-prey encounters) are social in some broad sense. They all help to define relations among individuals and within groups. Even hostility should be considered social.

Why so many nectarivorous birds are so unfriendly to one another is a good question. In my experience, they certainly are more actively aggressive on the average than many other birds of different feeding habits. One would suppose that the answer to this question must have something to do with the distribution and availability of food in time and space. In many cases, nectar renews itself, is secreted again and again, sometimes daily or perhaps even hourly. There must be selection pressure upon hopeful would-be consumers to attend and defend such rapidly renewing resources as frequently and as strongly as possible. Attacks can be useful in this connection. So can opportunistic "traplining," jumping from site to site (see Lyon and Chader 1971; Wolf and Hainsworth 1971; Wolf et al. 1972; and Colwell 1973). This is competition on the qui vive.

Although hostility is widespread, it is expressed in different ways in different places. The humid cold zone of the Andes has its own topographic and geographic parameters. It is generally broad, thick (hundreds of kilometers across), and continuous toward the center of the region. It becomes narrow and then fragmentary, a series of ecological islands and narrow peninsulas, to both the north and the south. In the center, individuals of different species of the Diglossa cluster usually are separated from one another, at any given moment in time, by an elaborate system of mutual-avoidance patterns within overlapping territories. Avoidance can be acoustic as well as spatial. Individuals of different species apparently try to avoid singing simultaneously. (The situation is quite the reverse with rivals of the same species. These often sing at the same time, perhaps to intimidate one another and/or to confuse or swamp the acoustic messages of rivals. See Morton 1986.) On the outskirts or extremities of the cold humid zone, in the islands and peninsulas, the different species of the cluster tend to be mutually exclusive with nonoverlapping territories, presumably as a long-term consequence of actual fighting or at least effective threat.

Obviously, the resources in dispute are not critical enough, at any given time, to justify suicide attacks or other forms of extreme recklessness. They are, however, important enough to provoke a great deal of hostile behavior, some of it high intensity. But why two different kinds of hostile behavior? Why avoid in some areas and exclude in others?

Again the probable answer is simple in principle. Successes and failures of alternative strategies must depend, in part, upon frequencies of encounters and interactions. For members of the Diglossa cluster of the high Andes, the sizes and perhaps the shapes of patches of suitable habitat must be significant, even determinant of the tactics chosen. The geographical parameters control the numbers and probably the nature of the potential competitors likely to be encountered. Numbers are important.

Consider the possible sequence of events during many disputes among birds that actually face or fight with one another. Let us imagine, for the sake of simplicity, that a dispute begins with only two individuals involved. One of these individuals may "win" eventually. The loser will retreat and perhaps leave the area. This does not always solve the winner's problem. There may be other potential competitors in the neighborhood who will come forward in turn. The original winner may have to face and attempt to repel a series of opponents. Every close encounter is a risk as well as an opportunity. (The birds are small. So are their offensive weapons. They can still be dangerous on their own scale. The bills of *Diglossa* spp. are sharply hooked. Some hummingbirds can use their long and pointed bills for stabbing.) Increasing numbers of fights may accumulate risks without necessarily providing more or different opportunities. The same area may be fought over every time.

When and if densities are roughly comparable, as seems to be the case with many populations of the Diglossa cluster, series of violent encounters with competitors would be expected to be longer, more frequently repeated, on the average, in large areas than in small ones, simply because competitors are likely to come in an unending stream in a large area. An inhabitant of a small area, by contrast, may dispose of its smaller number of potential competitors, at least temporarily, by a relatively small number of fights.

The situation probably is (even) more complicated in actual fact. Referring to another bird of another region, the willow ptarmigan, *Lagopus lagopus,* Eason and Hannon (1994) make a suggestive comment: "Territorial defense against new neighbors appears to require greater expenditure of both time and effort than did defense against former neighbors." If the same general principle applies to members of the Diglossa cluster—as seems only reasonable—then the birds of the small extremities, the islands and peninsulas, are expending more of their capital per dispute than their colleagues of the large central block, but they probably are more than compensated by winning more decisively and having fewer disputes overall.

Trade-offs between time and energy can be various during most social interactions, hostile or nonhostile.

Even with complications, it seems eminently worthwhile for an inhabitant of a small area to take the risk of (fairly mild) fighting. The risk seems to be too great for an inhabitant of a large area. This individual has to become tolerant. Toleration is not necessarily cheap; but the more efficient the toleration, the better. Mutual avoidance seems to be reasonably efficient. The moral is clear: "If you can't lick them, you must put up with them. You should try to do so quietly and politely."

There is a more radical solution: "If you can't lick them, you should try to join them." This cannot be easy, but it has been attempted, apparently successfully, by many mammals and birds, including at least one other cluster of species in the Andes.

Before proceeding further, it may be useful to add a couple of points. One is a reminder, the other is an explanation.

The reader should remember that the particular avoidance and exclusion patterns cited above are *inter*specific. The same birds would react differently to individuals of their own species. There would be less tolerance or moderation because the individuals use the same resources in the same way *and* compete reproductively.

Full-blown avoidance or exclusion performances can become physically elaborate, with many vocalizations and other displays. They are discussed here rather than later in the book because the "basic" approach and withdrawal movements are themselves simple in form.

Kingfishers and Rollers: More Interspecific Behavior

The coraciiforms have already been introduced. Two types or groups are particularly relevant here. They are the kingfishers (taxonomic family Alcedinidae) and rollers (Coraciidae). The names are misleading. Not all alcedinids fish; nor do all coraciids roll. Still, the names seem to be embedded in the ornithological literature of the English language. There are resemblances among these birds. Not only are they predators; many of them hunt in the same ways. They tend to perch high, to "wait and see," and then pounce down upon their usually large prey.

Beachley et al. (1995) have recently discussed the theoretical economics of what they call "sit-and-wait foraging," especially the sitting parts. They cite, as examples, a few animals: caddis larvae, marine plankton, spiders, lizards. They do not cite coraciiform birds, but many kingfishers and rollers may be comparable. They are not perfect examples; they do not conform to the models of Beachley and his colleagues in all respects. Thus, for instance, they sometimes change perches frequently. It is also doubtful that their "presence does not reduce the future quality of the resource," in this case, food. They may have various effects upon future quality.

Species of *Coracias* take their prey on land, often from the ground or from low vegetation. Many kingfishers of the genus *Halcyon* and related forms such as the kookaburras *(Dacelo)* also take all or most of their prey on land. They are called *martin-chasseurs* in French. The species that take their prey from the water are called *martin-pêcheurs*. As a group, kingfishers seem to be more specialized, morphologically, than rollers. They also are more speciose. This is most notable in "Malesia," the region including southeast Asia, the East Indian archipelago, New Guinea, and adjacent islands, the probable center of evolution or diversification of the family. It might be fair to say that kingfishers of the *chasseur* type are, ecologically, a new improved version of rollers of the *Coracias* type. Certainly representatives of the two types are potential competitors: they use similar nestholes and take similar prey.

I observed five species of *Halcyon* and three species of *Coracias* in the Casamance region in the southern part of the West African republic of

Sénégal (Moynihan 1987a, 1988, 1990). The territories of most of the local kingfishers and rollers are widely overlapping with territories of one or more of their potentially serious competitors of other species (*widely* may be a significant term; see below).

Interestingly enough, individuals of different species of kingfishers and rollers manage to finesse face-to-face contacts by personal avoidance, both spatial and to some extent vocal, in much the same ways as do members of the Diglossa cluster in the central part of the Andes. This is not necessarily what would have been expected. Nectarivory and carnivory are different. Large-prey items are relatively scarce and scattered. When found, they should be very much worth fighting for. And yet this does not occur. Why? The really extremely formidable weapons of the species must be a consideration. It is also possible that the local kingfishers and rollers face one of the same problems as do the members of the central Diglossa cluster. The environment of the Casamance is "sub-Guinean," an irregular jumble or mosaic of forest patches, secondary woodlands, and open brush savanna. Similar habitats stretch far away to the north and east for many hundreds or even thousands of kilometers. Thus the birds of the Casamance might face an endless succession of intruders, too many to fight without risking unacceptable losses. Another factor may be minor but not irrelevant. The territories of the local kingfishers and rollers are relatively large, much larger than those of flower-piercers or most other nectarivorous birds. The size of the territories may help to reduce the frequency of accidental encounters. So again toleration has had to be accepted as a compromise faute de mieux.

Not all kingfishers and rollers are the same. Here I may cite my observations from other parts of the world.

The Indian region has fewer coraciiforms than West Africa. There is only one species of *Coracias* over most of the subcontinent and not many species of *Halcyon*. Where individuals are rare, social relations among the local species may be obscure, difficult for a human observer to ascertain. There are known to be wide spatial overlaps with mutual avoidance among certain species and populations. This is true, for instance, of *Halcyon smyrnensis* and *H. capensis* in central Nepal (pers. obs.).

A taxonomic note may be inserted here. *H. capensis,* the stork-billed kingfisher, used to be placed in a separate genus, *Pelargopsis,* because of its very large size and correlated allometric features. Fry (1980) merged the two taxa, probably correctly. Still, the size difference between *capensis* and *smyrnensis* is impressive, much greater than the differences among any species of the same genus in Africa. This must, of course, have consequences.

There also is mutual exclusion at some times and in some places in India. The range of *smyrnensis* extends to Orissa in eastern India. There, it approaches two related species: *pileata,* comparable to *smyrnensis* in size, and *amauroptera,* a close relative of *capensis* and almost equally large. The relations among the three species are remarkable.

Exclusion can take different forms. Two types come to mind immediately. In some cases, individuals of different species occur in different, even if adjacent, habitats. In other cases, individuals of different species defend territories against one another in patches of the same habitat. Both kinds of separation can be neat.

In Orissa, *amauroptera* and *pileata* seem to be confined to stretches of mangrove along tidal creeks. *H. smyrnensis* keeps away from mangrove but is abundant in neighboring areas, often within eyeshot or earshot of its congeners. Note the different relations of *smyrnensis* with *amauroptera* and with *capensis.* The two species that inhabit mangrove have largely nonoverlapping, exclusive territories. In my experience, individuals of the large *amauroptera* usually retreat before approaches by the smaller *pileatus;* but the latter do not usually advance very far before turning back. The stretches of mangrove are narrow. They may be compared with the islands and peninsulas of the extremities of the humid cold zone of the Andes. Perhaps the local kingfishers of Orissa can also hope to dispose of their relatively few rivals without too much difficulty.

There are still other kinds of separation. Among birds, migrations can be significant. They remove potential competitors for appreciable periods of time. The whole scale is different from that of local avoidance. Some African coraciiforms are definitely nomadic, moving long distances but only within the Ethiopian zoogeographic realm (Curry-Lindahl 1981). The social effects or consequences of such movements have not been studied.

Other migrations can be larger or longer, between realms, most notably between southern Asia and the Palearctic in the case of coraciiforms. In India, the Eurasian roller, *Coracias garrulus,* impinges upon the native *C. benghalensis* in such areas as Kashmir for a few months of the year during the northern winter (Ali and Ripley 1970). When and where this occurs, the two species probably are rivals to some extent. In other areas and at other times, there can be no competition among them at all.

I observed a partial parallel to the Indian case in coastal mangrove on the west coast of the Malay peninsula near Kuala Selanger. Again, species of *Halcyon* were involved; again, *smyrnensis* and *pileata,* plus *chloris,* a widespread species ranging from East Africa to the islands of the South Pacific. As in

Orissa, the local *smyrnensis* stayed outside the mangrove. The local *chloris* occurred both within and without the mangrove. The local *pileata* appeared to be confined to the mangrove, in fact to the outward seaward faces. The territories of *chloris* and *pileata* were nonoverlapping. Stress may have been avoided or reduced by the fact that the *pileata* of this region, if not of Orissa, are highly migratory, spending part of the year in northern Asia (King and Dickinson 1975).

Competition and aggression may contribute to the mix of selection pressures in favor of migration. It is not known if they are often the immediate stimuli for the initiation of really long-distance movements.

Kingfishers of the *chasseur* type are both numerous and peculiarly varied in the Papuan region of New Guinea (Beehler et al. 1986; Bell 1981; Coates 1985). They show a maximum of "separateness," quantitative and qualitative distinctions, among themselves. Some species are conspicuously different in size, for example, the closely related and similarly colored *Halcyon torotoro* and *H. megarhyncha.* All or most species have different habitat preferences. Some species, for example, *H. sacra,* are migratory, others are sedentary residents. There are remarkable specializations of feeding habits and diel rhythms. The hook-billed kingfisher, *Melidora macrorhina,* seems to be thoroughly nocturnal. This is unique among kingfishers and very rare among coraciiforms in general. Only a few possible examples can be cited. A ground-roller of Madagascar, *Urataelornis chimaera,* has been said to be "surtout nocturnel et crépusculaire" (Milon et al. 1973, following Appert 1968). Bursts of late crepuscular activity are characteristic of the aerial nightjar-like broad-billed roller, *Eurystomus glaucurus,* in Sénégal (pers. obs.). Another remarkable form of New Guinea is the shovel-billed kingfisher, *Clytoceyx rex.* It is supposed to feed largely on earthworms, presumably by digging rather than pouncing. The species might be described as a *martin-creuseur.*

All in all, Papuan kingfishers would seem to have kept interspecific contacts and competition to a minimum in the very region in which they have radiated to a maximum. There can hardly fail to be a causal relation between the two phenomena. (It might be suggested, in passing, that the radiation of kingfishers is partly due to a paucity of competitors. The mammalian carnivores of the region were all marsupials before the arrival of man and dog. New Guinea lacks the varied civets and genets, Viverridae, and mongooses, Herpestidae, of Africa and southern Asia.)

Why should these differences occur when and where they do? The value and availability of resources must be important. Doubtless other factors are also involved, probably in different combinations at different places at different times. Human observers may find it difficult to separate or disentangle the individual and miscellaneous ingredients of causal mixes. Yet a few generalizations can be hazarded very tentatively. If some of the coraciiforms and members of the Diglossa cluster are representative samples, then one might conclude that competing and hostile species are more likely to exclude than to avoid one another whenever exclusion is physically possible without prohibitive expenditures of time and energy.

In other words, for these particular birds in their particular circumstances, it would appear that steady and simple but restrictive patterns have usually been preferred to complicated, inconstant but potentially flexible, arrangements to solve the problems of interspecific social relations.

REDIRECTION

The term *redirection* has been used in different senses by students of the behavior of humans and of other animals. Here it is used in what seems to have become the conventional ethological sense. It is said to occur when aggression released by one stimulus is vented upon some other organism or object apart from the stimulus. The classical example is provided by some elements in the "grass-pulling" of the herring gull, *Larus argentatus* (Tinbergen 1952a).

Redirection in this sense can be convenient as an outlet. It occurs almost everywhere. It does, however, have several actual or potential drawbacks. It is often relatively expensive of time and/or energy. It may fail to impress the original opponent. In any case, like escape, it usually takes a performer away from the resource in dispute.

Redirection itself needs to be controlled. The most common form of control is a careful choice of targets. The choice is wider for redirection than for most other attack patterns.

Again there are examples from coraciiforms. In the course of *intra*specific disputes, rollers and *chasseur*-type kingfishers frequently dash away from their opponents to attack other birds such as small passerines, doves, and plovers. None of these victims or targets compete ecologically with rollers or kingfishers to any substantial extent. They also never fight back when attacked by coraciiforms. Their weapons are inappropriate or too small to inflict damaging injuries upon their aggressors. In this context, vis-à-vis coraciiforms, they are both inoffensive and nonoffensive. As targets, they are nearly as "safe" as the grass favored by herring gulls.

Not everything is always so easy. Watts (1995b) says that female mountain gorillas sometimes redirect aggression upon other females of their troop. They do this relatively rarely, however; probably because the attacked individual fights back. As a general rule, it probably is unsafe to trifle with female gorillas, but the behavior described by Watts may reveal or illustrate a widespread correlation. A brief survey of the literature would seem to suggest that redirection between species, from stronger to weaker, is more common than redirection between individuals of the same species. Conspecifics are always likely to have the best or most dangerous defenses.

A NOTE ON COSTS

The advantages that might be obtained by escape, withdrawal, avoidance, exclusion, and redirection are easy enough to imagine in a general way. So are the disadvantages. The reader will have noticed, however, that certain disadvantages or constraints, that is, time and energy costs, have been passed over rather lightly in many of the preceding pages. Of course, observers can hardly fail to receive impressions and to draw tentative conclusions about these matters. Still, it would be nice to have more hard data.

OLFACTORY REGULATION

It has been said that the medium is the message. Actually, the same or very similar messages can be encoded in different media. This is certainly the case in hostile or potentially hostile circumstances. The term *message* is used here in the sense of W. J. Smith (1965, 1977) and many other students. In effect, it means "information content."

Most birds cannot smell very much. Most mammals can detect a great variety of odors. They also produce odoriferous substances as signals. The general subject is reviewed by Ewer (1968), Ralls (1971), Eisenberg and Kleiman (1972), R. E. Brown (1979), R. E. Brown and MacDonald (1985a, 1985b), and Ferkin et al. (1995). Carnivores have been studied particularly well in the field and in seminatural situations. See, in addition to the above, MacDonald (1985) and Gorman and Trowbridge (1989) and such monographs as Leyhausen (1956) on cats; Schaller (1972) on the African lion; Kruuk (1972) on the spotted hyena; Rasa (1973, 1977, 1985) on the dwarf mongoose; Mills (1989), Mills and Gorman (1987), and Mills et al. (1980) on the spotted and brown hyenas; and Caro (1994) on the cheetah.

Olfactory communication has its own qualities. In the words of Gorman and Trowbridge: "It can be used when visual or auditory signals are difficult to detect, for example at night, under the ground or in dense vegetation. Odors can be deposited in the ground as scent marks and thus provide a spatial and historical record of an individual's movement and behavior. As signals, scent marks have the important property of remaining active for long periods, even in the absence of their producer." This last factor may be the principal advantage of the system, especially for the regulation of hostility.

Olfactory patterns can encode a wide variety of different messages—hostile, friendly, sexual. They can be produced in a wide variety of ways. Urine and faeces have their own smells. Many mammals, perhaps most, have special scent glands, each producing its own special secretion or mix of secretions, each presumably encoding different information.

The fact that these patterns are varied in usage and functions has induced authors and students of the subject to propose a host of semiclassical and pseudoclassical names for different kinds of performances. Among them are such neologisms as *gonohyone* (to be distinguished from *gamohyone*), *odmichnione,* and *thorybone.* There are even—using a prefix common in the behavioral literature—*allomones* or *alloiohormones.* These and many other terms are listed in R. E. Brown and MacDonald (1985a). They are perhaps too ingenious for a general discussion.

As usual, the important point can be put briefly: All olfactory patterns say, at least, "I am what I am. I am—or was—here." When individuals live together or associate frequently enough to spread odors from one to another, the message may also say, "I belong to such-and-such a social group." (Whether "I can do no other" should also be inferred is uncertain in the case of nonhuman mammals. See discussions of cheating and mimicry below.)

Most, but not all, olfactory signals are primarily intraspecific.

The glands producing secretions for scent-marking are located in or on different parts of the body, sometimes several parts, in different species. Thus, for instance, Gurnell (1987) cites seven kinds of olfactory signals, from various glands, in holarctic tree squirrels, *Sciurus* and *Tamasciurus* spp., a taxonomic group that is not particularly highly specialized for communication by chemical means.

A closer and more precise example is provided by the dwarf mongoose, *Helogale undulata,* as described by Rasa (1985). Individuals of this species are highly gregarious among themselves, in effect "eusocial." They live in very closely integrated groups. Much of the cohesion of such groups depends upon olfactory communication. The basic elements of the system do not seem to be very numerous, but they are used copiously, almost semi-continuously. Both males and females have two types of scent glands, both of which produce secretions that are deposited in stereotyped ways. The secretions of anal glands seem to be personal "signatures." The secretions of cheek glands register emotions or states of motivation. Individuals also pay close and frequent attention to the smells of faeces. These are deposited by all group members in heaps or piles at discrete sites. They seem to mark the limits of group home ranges (the animals are not really territorial).

The distribution of scent glands on the body is not random. There probably are functional correlations. Thus, it is my impression that arboreal mammals tend to use sternal glands on the chest when (in the same circumstances that) terrestrial mammals use glands of the genito-anal region. This seems to be true, for instance, of Australian "possums" or phalangers (Biggins 1984) and some New World primates, family Cebidae (Moynihan 1966, 1970a). The difference seems to be adaptive and practical; a matter of balancing or equilibrating in different ways to mark different substrates.

Many deposits seem to be more or less hostile, threats or warnings, "keep off" signs, as in dwarf mongooses. Many signals must be graded. Richardson (1991), speaking of an aberrant hyenid, *Proteles cristatus,* suggests, "The greater the territory size, and hence the difficulty of monitoring intrusions, the greater the threat transmitted by scent marks needs to be." (Of course, similar ratios between area size and strength of signal must apply to many different interactions in different media in different circumstances.)

One wonders if olfactory signals of a hostile nature could also have sexual valence. Certainly the deposits of pastes or faeces often are well placed to attract potential mates or sexual partners as well as to repel potential rivals. If the same deposits should indeed have this bivalent function, then they would be the analogues of many bird "songs." As far as I know, however, this possibility has not been explored by behaviorists.

Some areas of partial ignorance are somewhat surprising. Thus, there seem to be few published accounts of the *initial* formation of sexual bonds in the East African hyenas and cheetahs, which have been studied at such length in such favorable conditions (see van Lawick and van Lawick-Goodall 1970 in addition to the references cited above).

The costs of olfactory patterns are difficult to estimate. Some secretions seem to be chemically complex. Many mammals produce several kinds of secretions. Some or all of these may be metabolically expensive. The actual application of special secretions, the postures and movements, are usually relatively simple in hostile situations. But not always. The brown hyena, for instance, has to do some rather delicate maneuvering to deposit two different secretions on the same blade of grass almost simultaneously. There is perhaps a hint in the literature that the movements associated with hostile marking are slightly simpler in form, on the average, than those associated with nonhostile marking (Biggins 1984; also see Schultze-Westrum 1965).

Olfactory communication must be old in the history of mammals. Everything that we know or can infer of the biology of primitive Mesozoic mammals (references in Lillegraven et al. 1979; and Sigogneau-Russell 1991) would suggest that smell and touch were their most important senses. The proliferation of visual and acoustic signals in many modern mammals is a comparatively recent development.

As noted above, an advantage of some olfactory signals is that they do not need to be delivered face-to-face. This must obviate actual fights in many cases. There are, however, possible drawbacks (as usual). The encoding or arrangement of information may well be neat and efficient, but there could be difficulties with delivery—perhaps two kinds of problems. First of all, there might be temporal depreciation. Messages could become faint or blurred with the passage of time. Secondly—and partly conversely—olfactory signals, especially those distributed on substrates, must continue for at least some period of time, even if at diminishing strength and clarity. They cannot be shut down or turned off immediately should they become inappropriate or dangerous as conditions change.

The other ancient system, tactile communication, will not be considered here. There may be nuzzling or rubbing among individuals in the course of some forms of olfactory communication; but further comment will be deferred until the discussion of allopreening, allogrooming, and related behavior in chapter 3.

Before proceeding to a discussion of more complicated devices, it is convenient to consider or reconsider certain technical points concerned with signification as well as significance—that is, to review some principles of communication as they apply to the social control of aggression.

2
ASPECTS OF COMMUNICATION

A process that can be called *communication* is an important factor in social interactions. The term has been used in many different ways; some of these are listed in Morton and Page (1992). Here the term is used to mean the conveyance of information from one individual to another.

The classic discussion of the general subject, mostly as illustrated by humans, is in de Saussure (the edition before me is 1986). A very useful account of the communication process among nonhuman animals, and especially birds, is by W. J. Smith (1977). There are further comments and references in such papers as Dawkins and Krebs (1978), Lazarus (1982), and Moynihan (1991).

Smith suggests that "information is a property of entities and events that makes their characteristics predictable to individuals with suitable sensory equipment." This suggestion is almost certainly correct as a general rule; but the concept of predictability is sometimes difficult to handle. It can, in fact, be very tricky indeed, and in several senses.

There have been objections to the use of the term *information* in discussions and analyses of communication on other grounds, most notably by Morton and Page (1992). Part of the apparent problem may be semantic. Morton and Page believe that the term has implications about such matters as equality of benefits and alternative roles. In my opinion, these are *not* really integral to the definition as given above. I will stick with the Smithian statement. It means what it says, nothing more nor less. However defined, the transmission of information from one individual to another, in some way or another, must be as old and widespread a phenomenon as aggression, if not competition itself.

The instruments of conveyance, the behavior patterns that encode information, are signs, usually called *signals* in zoological accounts. There is an enormous literature on the "science" of signs, alternatively called semiotics *(semiotique)* or semiology *(semiologie)*. There are a number of philosophical, metaphysical, or metaphorical problems involved (see Sebeok, 1969, 1976; and Eco 1975, 1988). These rather high-flying questions (Foucault's pendulum?) will be largely ignored here.

I certainly will not attempt to consider the process of communication as a whole, or to reinvent the bicycle. Only a few points will be mentioned in the hope that they may help to focus attention on some aspects of the evolution of social controls.

RITUALIZATION

Almost anything and everything is potentially capable of conveying information and therefore of acting as a signal, even when communication is not the primary function. Thus, a bird building (constructing or excavating) a nest to accommodate its eggs and young may not, in most cases, be concerned to transmit a message. Yet an onlooker watching the activity can hardly fail to draw his or her conclusions as to the significance of the performance. Much the same can be said of many other activities such as hunting, feeding, drinking, preening, scratching, resting, etc.

Patterns whose communication effects are in some sense "incidental" to their other functions are often said to be unritualized. They can be compared, partly contrasted, with ritualized performances. These latter are the patterns that *have* become specialized in form and/or frequency *expressly* to convey information, and perhaps in some cases to do nothing else. Ritualized patterns usually are called displays.

All acoustic performances of all animals (apart from "incidental" noises such as rustling and other locomotory sounds) may be assumed to be displays in this sense (fig. 1). At least, they should be operationally treated as such. They may, in addition to conveying information, also provide "emotional relief" for internal stress, but this function seems to be essentially additional

Fig. 1. Ritualized display postures accompanying aggressive vocalizations by southern great skuas, *(Stercorarius skua)*. From Moynihan 1962a, 3, fig. 2.

or ancillary. I do not know of any specially produced sounds that are not specifically adapted to encode and transmit information.

The distinction between ritualized and unritualized is not always clear and neat with other kinds of signals. In the cases of postures and movements, for instance, there are intergrades and intermediates, if only because ritualized forms have been, and still are being, derived from unritualized forms in the course of evolution. (The reverse may also occur, but the possibility has received little attention from investigators.) Thus, it is permissible and sometimes convenient, in some contexts, to talk of slightly, partly, or highly ritualized performances (fig. 2).

Fig. 2. Hostile postures of large white-headed gulls. Right, a very aggressive form by the kelp gull *(Larus dominicanus)*. Left, a less extreme form by a Belcher's gull, *Larus belcheri*. From Moynihan 1962a, 150, fig. 24.

Acoustic repertories include many different calls and other vocalizations and even instrumental patterns such as the (very different) "drumming" performances of some primates (Kortlandt 1972), gallinaceous birds (references in Moynihan 1995), and woodpeckers (Short 1982). The vocalizations of terrestrial vertebrates may have begun, rather humbly, as heavy breathing by a primitive amphibian of the Paleozoic (Spurway and Haldane 1953). They have certainly proliferated and diversified since then.

Vocalizations have also been studied in fine detail and at almost infinite length. Think of the size of the literature on bird song alone! Here we are concerned with only a few functional aspects. Many sounds express some degree of aggression, many others serve to discourage or deflect aggression. This may be simply taken for granted. Some acoustic characteristics such as tonal quality, grading, discreteness, and perhaps even redundancy seem to be partly regulated by certain widespread structural trends. Presumably, these trends reflect motivation as well as ecological factors (see Morton 1977, 1982, 1994; J. L. Smith and Yu 1992).

It is perhaps worth mentioning that some of the general as well as particular qualities, advantages, and disadvantages of acoustic communication are very different from those of olfactory communication. Sounds do not linger on, nor project into the future, after a performer moves on. They shut down or turn off immediately when they stop. The memory may linger on, but not the signal itself.

Visual patterns must operate within the same constraints. So do the electric pulses that contribute to communication among fishes (Heiligenberg 1991).

Most of the movements cited in chapter 1 are either unritualized or only slightly ritualized (in orientation) per se, although some of them are often combined or associated with displays such as scent-marking and (precisely) vocalizations.

Costs must be different for different kinds of ritualized patterns.

Consider sounds. Thorpe (1972), partly following Hockett (1960a, 1960b) and Hockett and Altmann (1968), suggested that vocalizations have two great advantages as signals. When and if they are confined to vibrations of the vocal cords, they leave the rest of the body free to perform other activities. This is fine when the other activities are either "maintenance" or nonstrenuous. It has also been assumed that the vocalizations of birds can be produced without expending much physical or physiological energy. This may hold for simple calls and notes (e.g., Horn et al. 1995). It is not likely to be true of all vocalizations, much less of all audible performances. Some birds sing very loudly and for long periods of time with every appearance of effort. (Also see Eberhardt 1994 and Catchpole and Slater 1995.) Other references, partly contradictory, are listed in Gaunt et al. (1996).

The loud and complicated vocalizations of certain mammals, for example, some cetaceans (Payne and McVay 1971), spotted hyenas (references above), and primates (Carpenter 1934; Moynihan 1966; Tembrock 1974; Marshall and Marshall 1976; Tenaza 1976; J. G. Robinson 1979; Cowlishaw 1992), must consume very appreciable amounts of energy (fig. 3). So might the distinctive, independently derived, calls and choruses of anuran amphibians (references in K. D. Wells 1977; Rand 1988; and Rand and Ryan 1981). Audible performances that entail movements of other parts of the body, such as the drumming of vertebrates already cited and the stridulation of tettigoniid orthopterans (Pierce 1948), can hardly fail to be at least moderately expensive.

Then there are the long and elaborate sequences of hostile and/or courtship performances of birds and mammals that involve a variety of both acoustic and visual patterns. Some of these performances were early favorites of early ethologists. Their studies produced many publications. Among them are descriptions, discussions, surveys, and summaries (the last mostly by Johnsgard) of such birds as Anseriformes (Lorenz 1941, 1951–53; McKinney

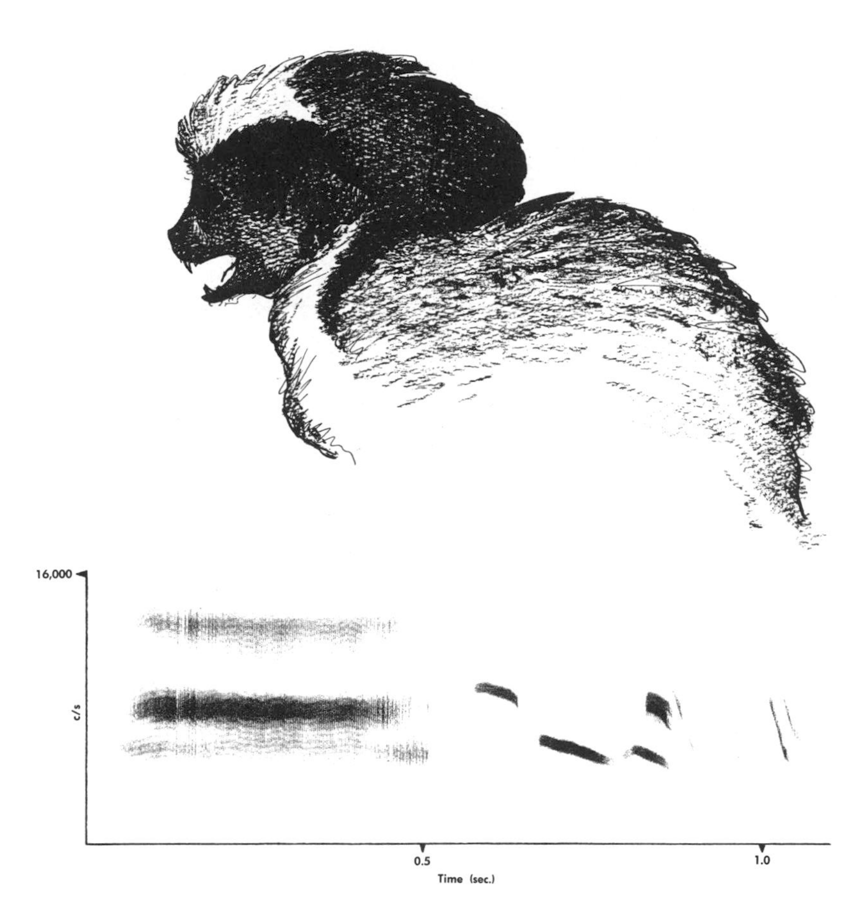

Fig. 3. Two complex vocalizations of New World monkeys. Above: Young *Saguinus geoffroyi* uttering infantile squeaks. From Moynihan 1970a, 61, fig. 21. Sonogram of an "infantile rasp" followed by "infantile squeaks" uttered by a young *Saguinus geoffroyi.* From Moynihan 1976, 167, fig. 45. Facing page: An extreme arch posture by an adult titi monkey *(Callicebus moloch)* with protrusion of lips and pilo-erection. From Moynihan 1966, 88, fig. 3. Sonogram of an "operatic" performance by an adult titi monkey *(Callicebus moloch).* From Moynihan 1976, 159, figs. 29 and 30.

1961; Johnsgard 1965), Pelicaniformes (Kortlandt 1940; van Tets 1965; Nelson 1978), Galliformes (Kruijt 1964; Johnsgard 1973, 1986, in addition to previous references; Moynihan 1995), Gruiformes (again Johnsgard 1986), and Lari (Tinbergen 1960; Moynihan 1962a; Furness 1987). There are corresponding studies of mammals such as elephant seals (Le Boeuf 1985), bats (e.g., Bradbury 1981), carnivores (references above), and ungulates (Walther

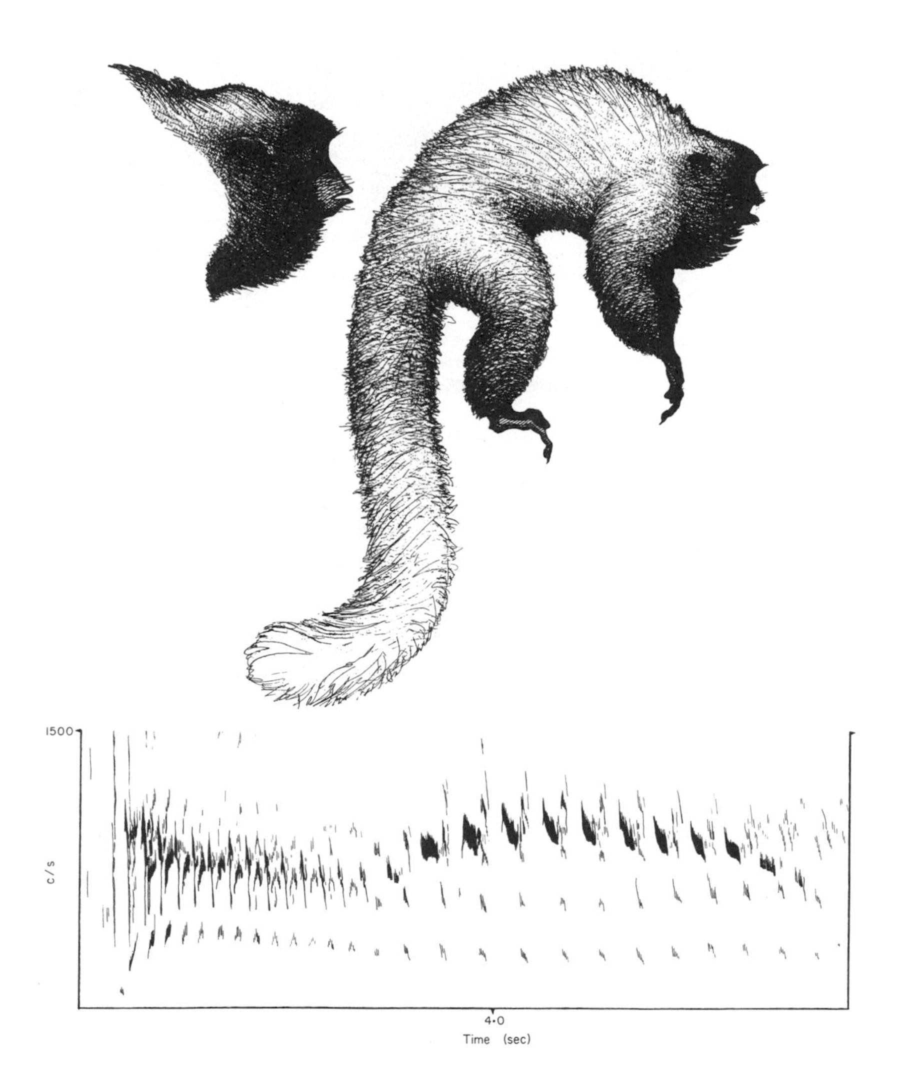

1984; also such individual monographs as Schenkel and Schenkel-Hulliger 1969; Schaller 1977; Gauthier-Pilters and Dagg 1981; Sowls 1984; Berger 1986; Rubenstein 1986; Schilder 1990). All this activity must be very costly; perhaps even more so in time than in energy.

Coleoid cephalopods were mentioned above. The species of this group that live in well-lighted waters have elaborate visual signal systems of remarkable, nearly unique, properties. They show many different color patterns. They change these patterns in different ways to produce hundreds, perhaps thousands, of combinations and permutations (fig. 4). The mechanisms are complex. The animals have elastic cells containing pigments in

Fig. 4. Some complex patterns of young Caribbean reef squids *(Sepioteuthis sepioidea).* In this case or these circumstances, the animals are being cryptic against sargassum weed. From Moynihan 1985, 50, fig. 22.

their skins. The pigments are various. The cells are under independent nervous control. They can be expanded or allowed to contract, singly or in groups, very rapidly. The different colors and combinations can signal aggression as well as other social reactions.

The brains of several species, the "common" European octopus and the "common" European cuttlefish, have been very well studied (J. Z. Young 1977; and M. J. Wells 1978), but little attention has been paid to the costs of the color systems. It would be interesting to know more about the bioluminescence of deepwater cephalopods (R. E. Young 1981) and "fireflies" (Lloyd 1983).

In the absence of adequate cost/benefit analyses, one is reduced to educated guesses. My "intuitive" feeling is that ritualized signal patterns of most birds and mammals are more expensive on the average, or are expensive more frequently, than most of their unritualized social patterns.

Some costs appear to be managed in peculiar ways. The small cuttlefish or cuttlefish-like squid, *Euprymna scolopes,* which extends into very shallow water, is known to flash a bright greenish light when disturbed (Moynihan 1983a). This must be a signal (among other things). The light is produced by symbiotic bacteria (Herring 1977). From the point of view of the "host," this display must be very cheap indeed. Free icing on the cake.

A few more notes on displays: Unlike escape and many types of redirection, they can often be performed in place. This is useful. Some correlates are less fortunate. Displays can be distracting to the performers themselves. They can be only too obvious, attractive to potential predators. A good example is provided by Ryan (1985): the vocalizations of male Tungara frogs *(Physolaemus pustulosus),* designed to entice females, expose the callers to attacks by predatory bats.

What you gain on the swings, you lose on the roundabouts. This may be a widespread principle of social life.

PREDICTABILITY

It must be assumed, as a preliminary working hypothesis, that any and every pattern, in itself (considered as a distinct unit by itself), always encodes or makes available the same information whenever it is performed. This is fully consistent with the (admittedly incomplete) evidence at our disposal. To suppose anything else would raise logical problems. There seem to be exceptions to the general rule, but they are very occasional at best, presumably usually brief, occurring only during certain evolutionary changes (see chapter 5). Yet it is obvious, on the briefest examination, that the same pattern can elicit, and often does, what appear to be different responses at different times. What is going on?

Most of the variation in apparent responsiveness must be "circumstantial." The same information is not likely to have the same effects in different circumstances.

Some of the most interesting effects are personal. All communication is directional. In most of the cases that concern us here, there are sender(s) of messages and receiver(s) of messages. A receiver may interpret the message of a sender correctly. When and if so, it should be able to predict the behavior of the sender *if and as long as* conditions do not change. Of course, the receiver can change conditions by changing its own behavior. The sender may change its behavior on its own initiative, or in response to the receiver. A receiver may fail to catch up with a change by the sender. And so on. Some prophecies (expectations) are not self-fulfilling but rather self-limiting.

Another possibility is that a receiver simply reads a message wrongly by mistake. This is rare during intraspecific encounters, but it must occur sometimes.

Factors of this kind might help to explain why Caryl (1979), using data from Stokes (1962a, 1962b) on blue tits, Dunham (1966) on rose-breasted grosbeaks, and Andersson (1976) on great skuas, found it difficult to identify the intentions of the performers of certain displays. Threat displays were the most problematic. It is true that they may tend to provoke a greater variety of responses than do many other patterns. Still, Caryl's results are unexpected; on the face of it, improbable. They need to be confirmed. Why should the same or similar information be treated so differently?

RELIABILITY

Confidence limits are debatable. At least, they have been debated.

Information can be virtually true or false. In all cases, signals are designed to induce responses. From the points of view of responders, some reactions may be correct, with good results, while others may be wrong, with bad results. Bluffing certainly does occur. All sorts of animals may swell up, increasing body size, or raise feathers or hair, crests and ruffs and frills, to increase apparent body size (figs. 5 and 6). This is sometimes effective in discouraging attacks by predators. It can also discourage rivals and competitors during hostile interactions among individuals of the same or similar species. The discouraging behavior might be called "braggadocio." When used discreetly, it conveys a message that is only a minor distortion of reality, a very little white lie.

Fig. 5. Hostile displays of pygmy marmosets, *Cebuella pygmosa*. Note the various kinds of pilo-erection, or ruffling. From Moynihan 1976, 43, fig. 7.

Fig. 6. Hostile postures of the grey gull *(Larus modestus)*. Ruffling is one of the tactics used to increase apparent body size. Top: a hunched posture with ruffling assumed by juvenile birds. Middle: ruffling superimposed upon a preattack posture. Bottom: A forward hunched posture with ruffling. From Moynihan 1962a, 61, fig. 8.

A similar and probably related phenomenon depends upon what has been called "the handicap principle." This was originally named by Zahavi, who discussed it in a number of papers (1975, 1977, 1987, 1991, among others). What the principle suggests is that an animal may carry or show off some character that might have been supposed to be burdensome, a handicap, simply in order to demonstrate or advertise that it is, in fact, strong and fit enough to carry the burden without suffering. Ninety-seven-pound weaklings need not apply. Ideas similar to those of Zahavi are widespread in the literature (e.g., Rohwer 1975; Dawkins and Krebs 1978; Enquist 1985; Maynard Smith and Harper 1988; Johnstone and Grafen 1993; Johnstone and Norris 1993; Petrie 1992; Møller 1994). The burdens are called "badges" in some of these papers. The badges of birds, the best studied group from this point of view, include conspicuous plumage marks (e.g., black bibs and iridescent ocells) and elongated tail feathers. The horns and hornlike structures of mammalian ungulates may also function as badges, among other things. The mineral-rich antlers of deer, first grown and then shed every year, must be very close to the limit in the ostentation of handicaps.

Zahavi developed his principle during an analysis of sexual selection. He considered burdens as sexual signals. Still, there is no reason that the same marks cannot be used as signals during hostile interactions. Zahavi has always insisted that the showing of burdens is honest. The demonstrators really are fit. It may be supposed, however, that a little braggadocio creeps in from time to time.

A few years ago, there were suggestions that some cheating could be much more serious than mild bluffing (e.g., Maynard Smith 1979, 1982; and Hinde 1981). This opinion was not accepted universally (comments in Moynihan 1982b). For a while, the subject was disputed. Recently, everything has settled down to a received opinion (Maynard Smith 1991, 1994). This was summarized in the 1991 paper: "Thus it has been shown, for a simple model, that honest signals must be costly if there is a conflict of interest between signaller and receiver, but that cost-free signals can be honest if there is no such conflict."

Actually, as far as I know, observers have not reported many kinds of serious cheating during encounters between vertebrate individuals of the same or similar species. Real lying begins when vulnerable prey are threatened by dangerous predators. Then the prey may develop all sorts of really "deceptive" patterns, crypses, camouflage, defensive mimicry, etc., specially designed to be unread or misread by predators (although usually quite clear to conspecifics of the performer). Interesting as it is, this deception has nothing to do with the social hostility discussed here, although some of the formal elements included may be similar.

IMPLICATIONS FOR THE ARGUMENT

This brief discussion of aspects of communication is inserted for a simple purpose: to persuade the reader that the patterns described below are "for real." In many cases, their physical forms, occurrences, and consequences are well attested. The data appear to be reliable. If and when they are accepted as such, then the explanations of functions, and therefore of selection pressures, should follow logically.

3
COMPLICATED DEVICES

This chapter is concerned with controls that appear to be more complicated in form, and sometimes usage, than most of the corresponding patterns cited in chapter 2. Like the latter, they canalize and/or discourage hostility in general and aggression in particular. Many of them are distinguished, however, by being obviously (or even more obviously) of ultimately nonhostile origins—in some cases parental or "familial," in other cases sexual. Such patterns may work, at their first appearances and repetitions, by distracting antagonists and/or positively stimulating nonhostile feelings or responses. In either case, this is "interference" of a sort, something like a blocking maneuver, and usually occurs at a high intensity of motivation.

This social blocking is not the same thing as the interference competition mentioned previously. In ecological interference, an organism reduces the ability of another to use resources. During social interference among behavior patterns, an activity precludes or transforms the performances of other activities.

As a precaution to avoid possible confusion, I will sometimes refer to social interferences as "diversions." They do, at least, appear to be engrossing distractions.

The patterns considered in this chapter are highly varied. The information encoded is unequal and heterogeneous. One might recognize a single classification according to morphological criteria. This would include phase changes and switches, possibly of several kinds. One could also distinguish certain functional aspects of the interfering patterns and/or of their predecessors. The two classifications, form and function, are not commensurable. Neither are they contradictory. I will try to refer to both whenever possible.

The sequence of patterns will continue to be empirical and pragmatic. As a rough guide, however, it may be noted that most of the patterns discussed are derived from partly "familial" origins, or are somehow related to such patterns, while those discussed for the rest of the chapter are derived from strictly sexual, even copulatory, origins.

It may be assumed that controls of hostility, like all other biological characters, can be improved, although not always in all ways at all times.

Controls may be added and multiplied, presumably with profit, but only up to a point. There seems to be an upper limit to the total number of kinds of displays that can be accommodated in the behavioral repertory of any given species. Comparisons among closely related species (e.g., gulls) show that displays seem to become "attenuated," less effective, with age and repetition. They do eventually disappear. They probably do so *pari passu* with their replacement by new displays, that is, other patterns newly ritualized or being ritualized (see Moynihan 1970b).

DISPLACEMENT PATTERNS: ÜBERSPRUNGBEWEGUNGEN

Displacement patterns have been described and partly analyzed by ethologists for a long time. The classical papers are Lorenz (1931, 1935, 1939) and Tinbergen (1940, 1951, 1952b). This group of patterns is as motley as any other. Perhaps the most noteworthy, if not the most significant, feature of all displacement patterns is that they were unexpected by human observers when first seen. They were supposed to be "out of context." The sources of the known patterns are diverse. Most of the visible aspects of avian performances are derived from such activities as feeding, cleaning, and nest-building, or various components thereof. They can be combined with other indicative patterns such as overt elements of hostility. Thus, for instance, a bird looking backward over its shoulder may be performing an intention movement of escape ("looking away" or "cut-off") or the beginning of displacement preening of the back or scapular feathers. Similarly, a bird pecking at the ground may be performing displacement feeding as well as redirected attack.

Looking away may be effective in several ways. It may be a cut-off. It removes the face, eyes, and bill of the performer from the view of an opponent. This may end the encounter or interaction by virtually hiding the releasing stimuli. Looking away also presents a vulnerable target, the back of the neck, to possible attack. The advertisement of vulnerability, in one form or another, is characteristic of several different kinds of controls. Taking a carefully controlled risk may, in appropriate circumstances, help to avoid more serious or disagreeable consequences in the future.

Some displacement performances are quite elaborate. The "choking" of the herring gull (Tinbergen 1952a) and other larids (Moynihan 1962a) is a good example (fig. 7). This combination of patterns occurs in ambivalent hostile-cum-sexual circumstances. It includes distinctive sounds, traces of redirected pecking, perhaps grass-pulling, and turning or scraping movements. The latter obviously resemble nest-building. They will be discussed in chapter 6 in connection with changes in motivation.

Single displacement patterns and many bouts of displacement tend to be brief. They are more or less expensive or economical according to the form or nature of the activity used as a source.

The frequency of displacements is different in different animals. It is not coincidental that the examples cited above are all taken from the behavior of birds. Ewer (1968) notes, "One of the most striking things that emerges from a study of mammalian threat behavior is that almost all of it can be traced to origins in intention movements of attack, defense or retreat and very little to displacement behaviour."

I might add that I never positively identified a displacement pattern in any of the species of coleoid cephalopods that I observed at some length in the field or in the laboratory: *Sepioteuthis sepioidea, Loligo plei, Euprymna scolopes,* and *Idiosepius pygmaeus* (references above; also Moynihan 1983a, 1983b). There were, however, some suggestive writhing movements of the arms, something like cleaning, during some ambivalent or ambiguous social encounters.

ALLOPREENING AND ALLOGROOMING

Among the various soothing or calming devices that may help to keep the peace among mammals—and that are usually very noticeable in groups and colonies—are allopreening and allogrooming. These are "transferred" maintenance activities (Poulin and Vickery 1995). One individual preens or grooms another instead of itself (see, for instance, the diverse accounts of

Lorenz 1952; Gurnell 1987; Heinrich 1989; McGrew 1992; and Mooring and Hart 1995). Naturally, such behavior may have directly practical usefulness. It may help to keep the plumage or pelage of a partner in good condition and remove ectoparasites. As a ritualized social signal, however, it may help to reduce hostility between partners. Almost any pattern that involves bodily contact, when and if it can be achieved (by no means always a foregone conclusion), would be expected to contribute to an increase of mutual trust and confidence.

The distribution of allopreening in birds is widespread but discontinuous, or "spotty." The behavior is common in many orders, but apparently rare or absent in others such as the Anseriformes and Galliformes (certainly I never saw it performed by any of the various pheasants that I have watched).

The Lari, gulls and terns, illustrate some of the spots. Most species of this rather large and diversified suborder do not perform allopreening. There is an exception, however: the noddies of the genus *Anoüs.* These birds use allopreening as an effectively sexual signal in precopulatory situations (Moynihan 1962a; figs. 8 and 9). This is noteworthy, because all the other Lari have a precopulatory display of very different form and source. The development of allopreening, its ritualization, must have been an independent specialization of the noddies within the evolutionary radiation of the Lari as a whole.

Fig. 7. "Choking" postures of Belcher's gulls. Top right, by adults. Bottom left, by a juvenile. From Moynihan 1962a, 154, fig. 26.

Fig. 8. Allopreening as a sexual signal in the adult white noddy *(Anolüs albus)*. From Moynihan 1962a, 255, fig. 41.

Allogrooming seems to be more consistently present in mammals. In fact, I do not know of any major group of the class in which it is absent. Of course, it may take peculiar forms. Thus, for instance, the "pacifistic" touching and other tactile contacts of fossorial Mediterranean mole rats, *Spalax*, in the Sahara (Nevo et al. 1992) might be a vestigial remnant of a once more elaborate allogrooming display.

A note may be added in passing. Almost any regularly accepted form of bodily contact can be socially reassuring. The contact is not always grooming or preening (see, for instance, fig. 14).

ALLOPARENTAL BEHAVIOR

The term *alloparental* is recent; but it is not my invention. Because ethologists use the prefix "allo" when referring to transferred behavior, presumably *alloparental* means taking care of other individuals, usually young, as if they

Fig. 9. Allopreening as a sexual signal in the black noddy *(Anolüs tenuirostris)*. From Moynihan 1962a, 236, fig. 37.

were one's own offspring, when, in fact, they are not. (Presumably, taking care of one's own real offspring should be called "autoparental" behavior.)

Some degree of alloparental behavior, in one form or another, occurs in many different kinds of gregarious birds and mammals. Helpers at the nest are only one example.

In many species of primates, mothers do not carry and nurture their infants all the time. The infants may be "borrowed" temporarily by other members of the mother's social group. They often are treated well by the borrower. Epple and Katz (1983) give references for marmosets and tamarins (also see Tardif 1994). Blaffer Hrdy (1977) gives references for many other species of monkeys. One of the principal advantages of this behavior is supposed to be that borrowers can learn by doing. They can practice to become good mothers or good fathers as the case may be, while the mother gets time off.

One side effect must be that different individuals become accustomed to one another. This can also encourage mutual trust and confidence.

An interesting case has been recently described by Heinsohn (1995). It involves a bird, the white-winged chough, *Corcorax melanorhamphus,* one of the Australian mud-nest builders. This is another cooperative breeder. It specializes in kidnapping. Members of one group entice away the young of a neighboring group. The young are then put to work (finding and delivering food, helping to repair the communal nest) by their kidnappers.

There is a slight difference in strategy between the examples. Borrowing primates may increase their own personal fitness. In doing so, they may also increase the "fitness," survival and success, of the group as a whole. The kidnapping choughs hope to increase the fitness of the group. In doing so, they may also help themselves individually.

ALLOINFANTILE BEHAVIOR

By extension, *alloinfantile behavior* should mean reacting to other individuals as if they were one's own parents when they are not. Presumably the converse is autoinfantile. Only a few possible examples of the allotype will be cited.

During a dispute between two adult titi monkeys, *Callicebus moloch* (not a very gregarious ceboid, nor as horrifying as its name might imply), the losing individual suddenly leapt on to the back of its opponent, grasped tightly, and stopped all expressions of overt hostility (Moynihan 1966). This was very reminiscent of the behavior of young animals seeking comfort or protection from a parent (also see Ewer 1968).

Pigeons and doves are known (by scientists) to be remarkably aggressive animals. Fruit doves, *Ptilinopus* spp., observed in captivity (Goodwin 1983), carry their aggressiveness over into courtship and premating sequences. If a bird is attacked by its partner, it may lower its head and push it under the aggressor's body, rather like a squab wanting to be brooded. The movement hides the head of the pusher. This may be useful because attacks are often focused on the head and eyes. Hiding is, of course, a common precaution against attack, especially in the absence of enough space for effective escape (also see the note on cut-off, above).

Other transposed or transferred elements of infantile behavior—perhaps both allo and auto—probably were involved in the evolution of some of the appeasement and soliciting performances that are discussed in a later section.

PLAY AND PLAY SIGNALS

Play might as well be mentioned here, because it is often supposed by the nonspecialist to be primarily an activity of young animals. The supposition is mistaken in many cases. Adults can join in vigorously. There are hints in the literature that carnivores may be more playful than most other animals (dwarf mongooses have already been noted).

The definition of play has been problematical. For our purposes here, it may be sufficient to quote Wilson (1975): "We know intuitively that play is a set of pleasurable activities, frequently but not always social in nature, that imitate the serious activities of life without consummating serious goals."

Full play as such is incompatible with real fighting and aggression, not only by this definition. But the two types of behavior can intergrade or, perhaps more frequently, occur in mixed-up jumbles together. Thus, for instance, the strikes and pounces and chases of rough-and-tumble play can suddenly, often in response to inadvertent or accidental stimuli, turn into the strikes and pounces and chases of dangerous hostile conflict (also see P. J. Sherman 1995). In Wilsonian terms, the nonserious can become serious.

It will often be in the interest of animals to keep play and fighting as distinct as possible. Signals have been evolved to help. Many Old World primates have special "play faces." They may indicate good will and inoffensive intentions (references below). There are other patterns. Bekoff (1995 and earlier papers) has studied play signals as "punctuation" during interactions among individuals of three types of canids: domestic dogs, wolves, and coyotes (usually, if not quite correctly, referred to as *Canis familiarus, C. lupus,* and *C. latrans*). A particular pattern, the bow, is often interjected into play sequences, especially when these also include elements of hostile origin. By performing a bow, an individual seems to be saying something like, "I want to play despite what I am going to do or what I just did—I still want to play." Apparently, this does tend to placate potential opponents.

There is a neat correlation. Infant coyotes are more aggressive than infant or adult dogs or infant wolves. They also signal intentions to play more clearly (frequently?) than either of the other two types.

GREETING, APPEASEMENT, AND SOLICITING

The term *greeting* is perhaps rather general. It seems to have been used in several senses.

One supposes that there must be low-intensity greetings. They have been little studied. We do not really know—have not yet identified securely—the animal equivalents of the human nod or a plain hello. Perhaps the (usually literally) nondescript bird sounds that have been referred to as "call" or "contact" notes or "conversational chatter" serve this function to some extent. Certainly, they are sometimes turned on or off in what would seem to be appropriate circumstances (Moynihan 1976).

It is remarkable, by the way, that little or nothing has been said about possible greetings among members of coalitions or alliances within groups. There seems to be a general, if not explicit, assumption that such relationships are formed early in life, in the nest or at a mother's knee.

What *have* interested ethologists are conspicuous ceremonies. An example has been described and the general subject discussed by East et al. (1993). Not surprisingly, the example is the spotted hyena. (Hyenids are almost as instructive as primates or coraciiforms.) Adult spotted hyenas are large and powerful predators as well as bone-crushing scavengers. They have the teeth and jaws that would be expected. They also are gregarious, living in groups that have been called clans. The internal organization of clans is fairly complicated. There are separate dominance hierarchies for males and for females; but all adult females are dominant over all males. Females have peculiar, obviously highly specialized external genitalia. These mimic the genitalia of the males. The clitoris is large, long, and erectile like a penis. The labia are fused into a structure that looks like a scrotum. East et al. suggest that this complex of distinctive anatomical features is a consequence of—an adaptation to—the special social requirements of the species.

Greetings "involve two hyenas standing parallel, facing in opposite directions. The hind leg nearest the animal being greeted is usually raised while each animal sniffs or licks the erect 'penis' and the anal scent gland of its partner." The process is usually initiated by subordinate individuals, both among adults and among cubs. This relation, with associated data, led East et al. to suggest that the greetings of the species are ritualized forms of submission that confirm asymmetries of status between the participants or partners. (Presumably it is the subordinate that is proffering the submission. What is the dominant doing? Presumably not only confirming her relative status but also "accepting" the submission? And with the same or similar patterns as the subordinate!).

East et al. also extrapolate. They suggest that all submissive greetings, when and if they occur, should be derived from behavior patterns of subordinates. Thus, in male-dominated societies, the sources should be juvenile and female. It is only in the female-dominated society of the spotted hyena that the source is male. Si non é vero é ben trovato.

None of the closest relatives of the spotted hyena, the brown and striped hyenas and the aardwolf, has a pseudopenis—nor, as far as I know, a penile display. The difference is interesting, and perhaps less enigmatic, in some respects, than comparable differences in other phyletic groups. Female spotted hyenas are really *very* aggressive indeed. The published literature would seem to imply that they are more aggressive, at least in extreme cases, than the females of the other species. If this is so, then one can understand why spotted hyenas might find it useful to signal submission in some particular emphatic or exaggerated manner.

The dominance hierarchy among female spotted hyenas also is very steeply graded. The social distance between the highest-ranking and lowest-ranking females is relatively enormous. It may be supposed that high rank is positively correlated with aggressiveness. According to Frank et al. (1995), the reproductive success of the highest-ranking females in a group is 2.5 times greater than that of all lower-ranking ones. Dominant females have preferential access to food; their offspring have greater survival rates; their daughters have higher social rank.

The context is destructive enough. The highly modified genitalia of the female often cause serious problems during birth. Again according to Frank et al. (1995), dystomia (abnormal labor) costs the average female 16.1 percent of her potential offspring.

The system would seem to be a remarkable demonstration of both the advantages of aggressive behavior and the disadvantages of some indirect correlates of aggression.

Appeasement is another term, like *greeting,* that has been used rather loosely. Probably, it should be applied to any behavior pattern that reduces an opponent's tendency to attack without, at the same time, greatly increasing its tendency to escape. The submissive behavior of the spotted hyena conforms to this definition. But the several terms are not precisely synonymous. Almost by definition, greetings occur when individuals join or rejoin one another. Appeasement can occur in wider ranges of social circumstances. And certainly some appeasement patterns look very different from penile display.

Take, for instance, the "begging" by adult birds of many species. Examples are the head-tossing of gulls and some terns and the wing-quivering of most passerines. These patterns are used by young birds to induce, even force, their parents to feed them, at first in the nest and then for some limited period after fledging. They also recur in later life, at which time they are used by adults, usually females, for "soliciting" in partly sexual encounters (fig. 10). Alloinfantility again? Of course, these encounters usually are somewhat ambivalent; they include elements of fear and aggression as well as sex; they are nervous-making. When performed by adults, soliciting may have several functions. At the beginning of a reproductive cycle, it may induce one adult to feed another, as a parent feeds a chick—all to the good for females that need to accumulate resources to

Fig. 10. Begging and soliciting in the brown noddy *(Anolüs stolidus)*. Top: a typical hunched posture of a juvenile bird. Middle: a typical hunched posture of an adult female during food-begging. Bottom: a typical hunched posture of an adult female immediately after copulation. From Moynihan 1962a, 220, fig. 36.

make eggs. Later in the cycle, begging and quivering patterns may release or stimulate copulation. Throughout, they seem to reduce probabilities of fighting between potential or actual mates.

As social signals, begging and soliciting would seem to be appeasement. They are less dramatic, less tense in appearance, than the penile displays of spotted hyenas. But then, birds are less formidable and less dangerous then hyenas. They probably can afford to take their appeasement with some degree of apparent self-confidence.

Given the fact that there are supposed to be "inherent" conflicts of interests between parents and young (e.g., Trivers 1974, 1985; Godfrey 1995; Redondo et al. 1992; and other earlier references), it is remarkable that derivatives of infantile behavior are so often successful as appeasement. This tells us something about the natural behavior of birds as parents. Aside from somewhat exceptional cases of "malign neglect," birds generally take good care of their young. They may even want extra ones (remember the kidnapping behavior of white-winged choughs).

Incidentally, begging is the pattern that was supplanted and replaced by allopreening during the evolution of noddies.

ALTRUISM

The subject of "altruism," once fashionable but then less so (Packer 1986a), is still recurrent in parts of the literature (Heal 1991; Enquist and Leimar 1993). It may not be directly relevant to controls of competition and aggression, but the term is used, sometimes loosely, in discussions of related topics. A few brief comments may not, therefore, be out of place here.

A recent paper (Connor 1995) defines *altruism* as "a costly act by individual A that benefits individual B." This is nearly acceptable, as far as it goes. But it does not go far enough. It may even be misleading. The term means something more in common speech; it usually implies some real sacrifice of inclusive fitness by the giver of the benefit—a real "Philip Sydney effect" (also see Maynard Smith 1991, 1994).

As it happens, Connor gives an inadvertent example of loose usage in his first paragraph. He considers the case of a zebra, in a herd, that detects a lion and then bolts away in escape. He says that the zebra is selfish if it bolts silently, and that it is altruistic if it utters an alarm cry as it goes. On the evidence available, this is an uncalled-for conclusion. The bolter could be helping itself by warning the other members of the herd. They might be either kin and/or useful cooperators. The alarm cry could also tell the predator that it has been noted. This might obviate attack. "Pursuit deterrence" signals are common and widely distributed among mammals (Moynihan 1970; Sherman 1977; Smythe 1977; Holley 1993).

Alarm sounds of one or several kinds probably are universal or nearly universal among birds and mammals. To my knowledge, only some populations of Eurasian badgers, *Meles meles,* have been supposed to have nothing of the sort (Kruuk 1989).

Fortunately, in another recent paper, Clutton-Brock and Parker (1995a) give clear definitions, in diagrammatic form, of true altruism (virtually as the term is used here) and such related or pertinent phenomena as mutualism, reciprocal altruism, selfishness, punishment, and spite. There should be no reason for further confusion.

The definitions by Clutton-Brock and Packard are included in a discussion of punishment, retaliatory aggression, in animal societies. They suggest that punishment is "used to establish and maintain dominance relationships, to discourage parasites and cheats, to discipline offspring or prospective sexual partners and to maintain cooperative behaviour."

I might add that it must be a convenient outlet, of practical value, for the residual aggressive tendencies that may be supposed to lurk underneath the placid surfaces of even the most friendly associations.

The concept of altruism seems to be useful in studies of insects (Hamilton 1972). It may be less so in studies of other animals. In actual fact, I do not know that anyone has been able to make a convincing case for the existence of such behavior, as a regular occurrence, in any species of bird or mammal.

NOTES ON INFANTICIDES

Infanticide is another subject that has been much discussed, although not always very evenly. Some aspects of the subject have drawn more attention than others. General discussions include Eisenberg (1981), Hoogland (1995), Moynihan (unpublished data), and many of the papers in the collections edited by Hausfater and Blaffer Hrdy (1984), J. L. Brown and Kikkawa (1987), and Parmigiani and vom Saal (1994).

Infanticide is the killing, active or passive, of young. It is important to understand because there appear to be few constraints on aggression. Several different kinds of killing can be recognized. They can be classified in different ways.

Hoogland recognizes four kinds of infanticide in black-tailed prairie dogs. The act can be performed by lactating females or female immigrants or invading males. Passersby may also kill abandoned litters.

Moynihan (unpublished data) suggests a classification with a more evolutionary focus. There are three categories: (1) the active killing of the offspring of potential mates; (2) the active killing of miscellaneous young, including some who probably are not the offspring of potential mates; (3) the killing—or letting die—of one's own offspring.

Most of the recent discussion of infanticide was stimulated by a famous, even notorious, book by Blaffer Hrdy (1977). The book is a description of the social life of the Hanuman langurs *(Presbytis entellus)*, leaf-eating monkeys of Abu in western (peninsular) Malaysia. The local population of the species is divided between "harems"—groups of one dominant male with several females and their young—and bands of unmated "bachelor" males. Occasionally, one of the outside males attacks, defeats, drives away, and replaces the dominant male of the harem group, becoming the new alpha male of the group in his turn. He is then likely to kill all the infants that were in the group before the takeover. The infants presumably were the offspring of the old and now superseded male. Blaffer Hrdy and many other biologists have generally supposed that the "bereaved" mothers survive, remain, recycle, and become sexually receptive again in a relatively short period of time. The new alpha male may then, therefore, sire the new round of infants more promptly that would otherwise have been the case.

Similar phenomena have been seen to occur in some, *not all,* other populations of Hanuman langurs, other species of langurs, and other monkeys (Kavanagh 1983). Even more significantly, the same kind of infanticide seems to be characteristic of very different mammals, for example, the lions *(Panthera leo)* of the Serengeti Plains of East Africa (studies of this population have been recently summarized by Pusey and Parker 1994).

Actually, the behavior is not confined to mammals. Nor is the flow of aggression between the sexes always in the same direction. Wattled jacanas *(Jacana jacana),* freshwater shorebirds of the Central American tropics, are polyandrous. Eggs are incubated, and the young are raised by the males. Emlen et al. (1989), using experimental techniques, induced females of the species to kill the chicks of other females in order to gain access to the "bereaved" males.

These examples are diverse enough to indicate that this sort of infanticide is perfectly "normal," although perhaps rare and of scattered occurrence in special circumstances (again, perhaps scarcity of some resources).

Since the practice has been selected for by evolution, it must have, or have had, certain advantages. Many students have assumed that the increase of inclusive fitness accruing to the winning, taking-over sex is advantage enough in itself. This may well be true; but there may be other ways of phrasing the same point.

Of course, because infanticide is a natural phenomenon, there also must be disadvantages involved. The principal disadvantage must be the loss of inclusive fitness by the bereaved parents. The takeovers themselves often entail fighting, with the usual risk of death for some individuals and crippling injuries for some of the survivors. Thus there may be a loss of personal fitness in addition to the loss of inclusive fitness.

Option 2, the killing of various young instead of, or as well as, the offspring of potential mates, must also have both advantages and disadvantages, the gains in inclusive fitness probably being slightly less on the average, and the losses of personal fitness perhaps also being slightly less.

Another set of problems is associated with both 1 and 2. All the excitement, vocalizations, and other activities that may occur during these kinds of infanticide can be disturbing to companions and collaborators. Members of a group may become frightened and dispersed. The same activities can attract the attention of predators. The disturbed and dispersed individuals may be distracted and less wary than usual. They may, therefore, be particularly vulnerable to predation.

The third option, the killing (or more usually) the letting die of one's own offspring, has not usually been associated with the preceding phenomena in discussions of infanticide per se. Yet it shows striking similarities in function, and probably in evolutionary history, as well as equally significant differences. It has been best described in birds such as diurnal raptors (e.g., L. Brown and Amadon 1968), seabirds (e.g., Murphy 1936a, 1936b; Nelson 1978; Reilly 1994; St. Clair et al. 1995; Anderson 1995), and a highly specialized coraciiform, the southern ground hornbill, *Bucorvus leadbeateri* (described in Kemp 1988).

The ultimate evolutionary cause of this phenomenon seems to be the same in all the well-known cases. A parent may find it worthwhile to be optimistic or even over optimistic. Healthy and successful animals are capable of producing more descendants than can possibly survive. This tends to be true in the short as well as the long run. Many individuals can produce more young in a single "effort" than they can raise except in an exceptionally favorable year or breeding season. As it happens, most years or breeding seasons are only ordinary, mediocre. Some birds may still find it worthwhile to lay wagers. They begin by producing as many young as possible. If the year or season turns out to be optimal, then all the young may survive until maturity. All to the good. If the year or season is less than optimal, then one or more offspring will die, directly or indirectly, from lack of nourishment or other support (protection). Too bad, but unavoidable in the circumstances. The relevant factors are discussed in such publications as Fujioka (1987) and Mock (1984a and 1984b).

The fact that most of the birds that have adopted this strategy are predators upon relatively large and/or mobile or nomadic prey probably is not coincidental. Such prey may be rare or frequently difficult to discover.

The immediate or final cause of the death of the young in some of these cases may well be attacks by siblings, as well as, or in addition to, starvation. Technically, such aggression may be called *siblicide*. But it is the parents that prepare and preside over the process. The setup is not accidental or fortuitous. Thus, for instance, the vulnerability of young boobies, *Sula* spp., tropical Pelecaniformes related to gannets, is partly dependent upon the size and shape of the nests—depressions in the ground—made by the parents (Anderson 1995). In the case of the southern ground hornbill, which normally lays a clutch of two eggs, the second egg is smaller and presumably weaker than the first.

When and if the surviving young go on to flourish, then their parents may be said, in retrospect, to have been provident.

Of course there is a price to be paid. The provident optimists must suffer a considerable loss of inclusive fitness in the short run. But they usually do not run the risk of any appreciable loss of personal fitness. They are not likely to be injured themselves. The neglected or abused young cannot fight back, or do not do so effectively. And the whole process can be done so quietly that neighbors are not likely to be disturbed, nor predators alerted or attracted.

The general advantage to be gained by infanticide can be rephrased. They can be said to include the removal, suppression, control, and sometimes replacement of competitors—more precisely, in the case of such animals as lions and Hanuman langurs, the actual or potential competitors of the performers themselves and/or of the performers' current or anticipated future offspring. In the case of animals such as boobies and southern ground hornbills, the point of the behavior is to eliminate competitors of the most promising offspring even (and precisely) at the expense of less promising ones.

In all or almost all cases, the method used is aggression—more naked, open, or direct attack than in many other circumstances. The behavior looks simple, if remarkable. It can also be associated with other patterns in obviously complex systems.

PSEUDOSEX AND ALLOSEX

The terms *allosex* and *pseudosex* are meant to be abstract, slightly noncommittal, without any accompanying aura of preconceptions. They are broadly inclusive in some ways; certainly *not* in others. As a matter of convenience, they can be applied to all behavior of sexual (e.g., copulatory) origin when performed (usually as signals) in relations and situations that are not directly or immediately sexual at the time of performance.

We have touched upon these matters earlier. The submissive penile display of spotted hyenas would appear to be a sort of sexual derivative, if a slightly peculiar one. It is also an example of the "opportunism" of evolution. There is another example, using similar material in a different way. Male squirrel monkeys, *Saimiri,* also have a display in which the penis (a real one) is erected quite conspicuously. At least as performed by adults, this seems to be an expression of male dominance, perhaps threat, but certainly not appeasement (DuMond 1968; also see Ploog and MacLean 1963; MacLean 1964; Ploog 1967).

Other animals and patterns should be cited: first some more primates, then some more birds. These examples are important in that they provide the details to show how these complicated sexual tactics are actually used by particular species.

Monkeys and Apes

Some monkeys and apes can perform a variety of sexual and sexually driven patterns that are not, at first glance, entirely "straight." Many of these activities have been lumped together under the general rubric of "homosexual." In a recent review of such behavior in nonhuman primates, Vasey (1995) defines homosexuality as "genital contact, genital manipulation or both between same sex individuals." This definition also is broadly inclusive. It is not, however, the same as the definition of pseudosexuality given above. The two categories are widely overlapping but not identical. As it happens, some of the pseudosexual patterns considered here are also homosexual; but others definitely are not. Pseudosex can occur among individuals of opposite sex as well as among individuals of the same sex.

In his review, Vasey notes that homosexual behavior, by his definition, has been recognized in thirty-three species of nonhuman primates. The distribution certainly is not random. Vasey also notes that this behavior is not entirely homogeneous. He considers possible functions, including, among others, dominance assertion, tension regulation (lowering), reconciliation, and alliance formation. The same functions can be subserved by pseudosexual patterns as well as strictly homosexual ones.

Ethologists have concentrated on one particular pattern, a form of pseudocopulation sometimes called "dominance mounting." There is a useful account of this behavior in Jolly (1972). She gives many references to classic older works such as Kummer (1968) and Wickler (1967). The physical aspects of the pattern are easy to describe. A dominant individual of either sex asserts its authority by mounting a subordinate, with or without pelvic thrusts. For the dominant individual, the behavior may be simply punishment for a lack of respect. For the subordinate individual, it can hardly fail to be appeasement. Of course, both individuals may also benefit if they can become accustomed to the close physical contact (see comments on allogrooming, etc., above).

A species that has been studied in considerable detail in recent years is the bonobo, or pygmy chimpanzee, *Pan paniscus.* Among the relevant accounts are de Waal (1988, 1995), Furuichi and Ihobe (1994), and the papers in Susman (1984). In his 1988 paper, de Waal notes, "Sociosexual mounting appears to be this species' most important mechanism of tension regulation; it serves as an alternative to aggression, as a peacemaking ritual." The general effect is supposed to be reassuring.

All this can be put in comparative perspective. Tree sloths do not, as far as I know, exhibit dominance mounting. Not surprising, in view of their nearly solitary habits. More remarkable is the fact that many species of primates also seem to lack this behavior. If it is so useful to some species, as does indeed seem to be the case, then why is it not present in the repertories of other species? According to Vasey, homosexual behavior, presumably including dominance mounting, is most widespread among Old World monkeys, Cereopithecidae, and apes (Pongidae, including Hylobatinae). Yet even here, it is supposed to be absent in such forms as *Macaca sylvana,* howler monkeys (*Alouatta* spp.), and gorillas. This spotty distribution would suggest that homosexual and/or pseudosexual patterns have been developed several times in the course of evolution—or perhaps lost several times.

On a broader taxonomic scale, it is noteworthy that homosexual behavior seems to be absent throughout the whole series of prosimians, the lorisoids and lemuroids. This in spite of the fact that some forms, for example, *Lemur* and *Propithecus,* are very monkeylike in other aspects of their biology. Presumably, prosimians and monkeys have similar social problems, but they seem to cope with them by using different behavioral devices. Another example of alternative solutions.

Some Genital Displays

Perhaps with cercopithecids in mind, Jolly (1972) is clear and amusing on the roles of the protagonists in dominance mounting: "It is an advantage both to 'fool' and to be 'fooled' in this situation." There are few limits to the ingenuity and ingenuousness of fooling around.

The adult females of many Papionini, macaques and baboons, have bare skin, usually colored pink or reddish, around the genito-anal area and on the buttocks. The males of some species have similar patches of exposed skin (Ewer 1968). The patches of both sexes are highly visible from the right (correct), rear, point of view. They appear to function as signals, perhaps appeasement, among other things. The patches of females swell at oestrus. They may, therefore, indicate receptivity to male approaches and help the males to find the correct orientation for copulation. All this is quite honest.

There can be complications, however. What might be called "skin displays" have become very elaborate in several forms (Morris and Morris 1966; Wickler, his 1963 paper as well as the 1967 one). The breasts and nipples of the female gelada baboon, *Theropithecus gelada,* have become bare and red and are surrounded by a border of white papillae. The arrangement resembles the female's genital region. Wickler thinks that the resemblance is not coincidental. It might be "automimicry." The face of the male mandrill, *Mandrillus sphinx* (again not the most appropriate of specific names), is marked by bright red and blue skin. The penis of the species is bright red; it is viewed against the background of a blue scrotal sac. Again, this is not likely to be coincidental. It is suggested that, in both species, the original signal of the genital area has been copied in a forward position in order to be easily visible from more points of view.

The functions of the mimic signals are not really well known. Perhaps appeasement in the female gelada. Perhaps threat in the male mandrill. In any case, the system is simple in one respect: the mimic cannot be separated from the model, or vice versa.

Sexual Dimorphism in Kingfishers

The evolution of sexual dimorphism is something of a problem (Lande 1980; Burley 1981; Björklund 1984). Animals may be said to be dimorphic when males and females are distinguished from one another by possession of different characters that are both permanent and noticeable. Discussions of the subject are usually concerned with visible features—as in the case of the pied kingfishers—although there is no theoretical reason why other kinds of characters might not be involved instead or in addition (see notes on olfactory and acoustic communication).

The pied kingfisher, *Ceryle rudis,* is widely distributed in tropical Africa and Asia. It occurs in several subspecies. The best-known populations are in East Africa, along the shores of Lake Victoria. They have been described by Douthwaite, Reyer, and others (references above). I myself have observed groups, individuals and family parties of the species in West Africa, in the lower Casamance region of southern Sénégal; in Tamil Nadu in southern

India; and in central Nepal, in the Royal Chitwan Park, at irregular intervals between late 1976 and early 1987 (see appendix 1).

As far as I know, the species is everywhere a classical *pêcheur* rather than a *chasseur.* Even as a kingfisher, it is notably aggressive. It also is comparatively large. The sexes are easily distinguished by sight. Males and females have different bands or patches of black on the white breast.

The problem of sexual differences of appearance may be particularly baffling in kingfishers. Burley seems to imply that dimorphism is primitive among birds in general. Among kingfishers, however, it is monomorphism that seems to be primitive. It is characteristic of the great majority of living forms. Only a few groups of species are dimorphic. These groups are diverse and probably not closely related to one another. The particular characters distinguishing the sexes are different in different groups and species. Thus, the various dimorphisms of these kingfishers probably are secondary developments. Why were they selected for? Ecological and adaptive parallels are not always obvious to a human observer. (Certainly, males and females do not seem to play more different roles or to occupy more separate niches in dimorphic species than in monomorphic ones.)

I can only suggest that the crucial factor is visibility among the birds themselves. How easily or *how quickly* can males and females recognize the sex of a potential interlocutor? In environments with good, clean, and unobstructed visibility, all or most of the local kingfishers are monomorphic. In these circumstances, each individual usually has time to recognize others gradually or at a distance. They do not need to rely upon conspicuous banners or pennants. These "banners" and "marks" are not the same as the "badges" mentioned earlier. They are supposed to identify sex, not to advertise fitness. When visibility is bad or obstructed, on the other hand, individuals may find themselves "precipitated" upon one another at short distances and brief notices. They may indeed need conspicuous marks for instant recognition. Simply put, for whatever the reason, there are more species of dimorphic kingfishers in dense vegetation, forest, and perhaps mangrove than in open or semiopen areas.

Evidently, dimorphism is not prerequisite to successful courtship and copulation; but it may help to avoid unnecessary fights between the sexes.

The monomorphic *Halcyon* kingfishers of mangrove in Orissa and peninsular Malaya would appear to be exceptional. Perhaps they are recent invaders of their present environment.

The aggressive behavior of pied kingfishers is controlled or canalized in different ways at different times in different places by a variety of social mechanisms. Geographic and temporal variations are, in fact, surprisingly large. The Lake Victoria populations of *rudis* are gregarious during breeding seasons. There is a skewed sex ratio. Associations of one female with five males have been reported. I found something comparable in the Casamance. During the breeding seasons, there were groups, even clans, of up to twelve individuals with one, two, or three females.

In the Casamance, individuals occurred in and around tidal areas and patches of mangrove. Although they sometimes flew over wide areas, they also were frequently crowded together on perches, perhaps only a few centimeters apart on occasion. Personal relations within clans were not always easy to follow in detail, but certain features were obvious. Any female could choose a particular male as preferred mate or consort. She allowed the favored male to copulate with her repeatedly. The same female could also associate with other males, less frequently or less closely than with the primary consort but nevertheless fairly regularly. Even the various secondary males seemed to be ranked in the female's eyes. At least some of them attempted copulation with the female. It was impossible for me to determine if insemination occurred. There was squabbling during these attempts. Still, the procedure appeared to be too ordinary or normal to be usefully described as forced (or, in current jargon, "forced extra-pair copulation").

Heterosexual relations may have been complicated. Homosexual relations were remarkable. Many males, all or most of them secondary or supernumerary, established something like a "buddy system" (a restricted brotherhood). They paired up with one another, two by two. The members of such pairs often allowed one another to land close by without much disturbance or real disputing. At other times, they performed brief, flurried patterns that looked like "mountings." These certainly were accompanied by vocalizations and other hostile displays. They were not, however, closely correlated with attempts to mount a female, that is, they were not carelessness or redirection.

The pied kingfishers of the Casamance live in a female-dominated society if there ever was one, but they do not seem to perform anything like a specifically masculine pattern to indicate submission or appeasement (unless copulation itself could be considered as such).

Presumably, the sexual (i.e., mounting) attempts of male *rudis,* although accompanied by overt hostility, may still reduce the intensity or violence of a dispute below the levels that would otherwise have been attained in the absence of such attempts. The parallel with some primates is striking.

Also presumably, many of the birds of the Casamance should be kin of one sort or another, but they may well have to accommodate immigrants from time to time.

There are other populations of *rudis* that seem to have simpler social structures. I never saw large groups of the species in India or Nepal, only pairs and small parties of three or four individuals (nuclear families?) along freshwater streams or by the shores of ponds or tanks.

Simplicity can be intermittent. In the Casamance, in the *same* areas where larger groups were seen in the breeding season, there were only single birds, pairs, and small parties left a few months later, after the end of the season. Moreover, the sex ratio of these remaining birds was approximately equal. I have no idea where the other, accessory, or supernumerary males may have gone.

The peculiarly lopsided clan structure of *rudis,* where and when it occurs, may be a consequence of crowding. There may also be some coincidence with flushes of prey.

It should be noted, in passing, that the formidable "giant" kingfishers of the closely related genus or subgenus *Megaceryle* live in large or dispersed territories and have essentially ordinary social behavior, in pairs and small families, throughout their ranges in Africa and the Western Hemisphere (Moynihan 1987 b; Davis 1985 and pers. comm.).

Blue-Bellied Rollers

I studied another coraciiform, the blue-bellied roller, *Coracias cyanogaster,* in the Casamance (Moynihan 1990). Thiollay (1985) studied it at Lamto in the Ivory Coast. The populations of the two regions have similar ecologies. The general or usual situation can be summarized briefly, and a partial quotation from the abstract of my paper is added. Blue-bellied rollers are *martin-chasseurs,* sit-and-wait pouncers. They differ from other *Coracias* species in preferring humid or semihumid areas. In West Africa, such areas "are naturally forested. Blue-bellied rollers often perch high in trees. Still, they prefer to get their prey as low as possible," on or near the ground. "This means, in effect, that they are dependent upon clearings [fig. 11]. Patches of bare earth and sparse vegetation probably were few and scattered before the spread of agriculture. In these earlier circumstances, blue-bellied rollers must have had to concentrate upon whatever open spaces were available. They may have been more crowded on a restricted local level than any other species of roller."

Despite general resemblances of ecology, the social arrangements of blue-bellied rollers may be different in the Casamance and in the Ivory Coast. I will begin with the former.

Fig. 11. Three postures of the blue-bellied roller *(Coracias cyanogaster).* Top: an individual scanning the neighborhood for prey. Middle: a forward-leaning, preflight posture. Bottom: looking downward before pecking at a substrate; the neck feathers are slightly ruffled. From Moynihan 1990, 2, fig. 1.

There seem to be clans within the local population, occasional large groups of up to twelve individuals. More commonly seen are pairs or groups of three or four. There is little sexual dimorphism in plumage. Males may be slightly larger than females; they are known to be slightly heavier at Lamto. On the data available, the blue-bellied rollers of both the Casamance and the Ivory Coast have approximately equal sex ratios. Typically, within groups of three or four birds, two individuals are firmly bonded to one another, while a third seems to be less closely attached, and a fourth, if present, may be almost semidetached.

Pairs and groups have usually nonoverlapping home ranges. These can be defended as territories. The forms of disputes are characteristic. When two parties are involved, each party has its own champion, perhaps the alpha male of its group. During low-to-moderate-intensity disputes, the champions tend to follow or chase one another, sometimes displaying vigorously and aggressively, at other times sitting deceptively meekly side by side. The other members of the two parties may simply watch at a distance.

This gladiatorial situation can change if a dispute heats up. A champion may suddenly fly away from its opponent and join its presumed mate among the spectators. It is not only looking for moral support. Its arrival may be followed by an outburst of copulatory-type behavior. One individual, usually the joiner, immediately mounts the other. It begins to beat its wings, utters a pumping vocalization, lowers its tail, and so on. The lower individual may raise its tail, as is typical in "real" copulations, with or without vocalizations.

Some encounters are simple. The mounter eventually slides off the mounted. Nothing more. Other interactions, considerably more numerous, include "reversals." The original mounter slides off and is mounted in turn by the individual that had been mounted earlier. To quote again: "First A on B; then B on A . . . both the individuals involved played both male and female roles in rapid succession." I saw reversals in at least seven pairs or groups. The numbers of reversals per "bout" were varied. At one extreme, I counted 24–28 mounts with 12–14 reversals. At the other extreme, I saw several cases of only two mounts with one reversal.

Simple mountings during disputes and multiple reversals occur throughout all or most of the year in the Casamance, while actual breeding, the laying of eggs and the rearing of young, is strictly and narrowly seasonal (Morel and Morel 1982).

The difference in timing is revealing. At least some of the mountings during disputes in the nonbreeding season must be pure display, "only" social signals. Given the circumstances, they probably are hostile.

Some possibilities can be discounted. These patterns are not likely to be used by individuals to discover their sexual identities or to sort out their future sexual roles, as has been supposed, rather dubiously, to be true of some partly similar-looking patterns of domestic pigeons, Eurasian cormorants, and several species of woodpeckers. Using sex to define sex would appear to be supererogatory.

Nor do the peculiar patterns of blue-bellied rollers seem to be examples of "mate-guarding" in the conventional sense of the term as used now. It is quite possible, of course, that mountings, even during disputes, do somehow reinforce the bonds between the mated birds performing the behavior.

After all, the physical contact is close (again, see comments on allopreening and allogrooming). But the mate-guarding that has been described and discussed in recent publications (e.g., Davies 1989, 1992; Petrie 1992; Whittingham et al. 1995) is a rather different thing. Most authors have been less concerned with long-term bonding than with the prevention of "extra-marital" affairs and the regulation of sperm competition. Presumably, sperm are not in play among blue-bellied rollers in the nonbreeding season. The most significant aspect of this behavior lies elsewhere.

Among many animals, from horseshoe crabs *(Limulus)* on up, and certainly among most birds and mammals, the sight of individuals copulating nearby is likely to provoke a forward rush of bystanders and passersby, apparently trying to break up the performance, or perhaps hoping to steal an extra, free, copulation in the general confusion. It is, therefore, very highly suggestive indeed that nothing of the sort seems to happen among the blue-bellied rollers of the Casamance. Here, the spectators of the mountings, although obviously hostile (witness their chasing and aggressive displays), do not intervene actively to attack the individuals apparently engaged in sex before their very eyes. In this case, the conspicuous demonstration seems to be positively off-putting or repellent to the onlookers—effectively, a kind of defensive threat. Whatever the performers may be doing to each other, they appear to be strongly influencing, controlling, the behavior of the spectators. This, presumably, is the point.

The Ivory Coast may provide a variation. The blue-bellied rollers of Lamto are supposed to be adventurous. Thiollay says that "up to three males may copulate successively, even two or three times, with the same female," and "one male may copulate with two females at an interval of ten minutes" (my translation). Again, apparent copulations occur well outside the breeding season. Why they should do so here is not yet evident.

None of the other species of *Coracias,* nor of the related genus *Eurystomus,* has been recorded as performing reverse mountings. The European *C. garrulus* is well known. I can vouch for the absence of reversals in at least some populations of some other species. I was able to observe, at considerable length, individuals of *C. abyssinica* and *E. glaucurus* in the Casamance, *C. benghalensis* in southern India, and *orientalis* in peninsular Malaysia. Briefer observations were made of *C. naevia* in Sénégal and *E. gularis* in Liberia.

Thus, the pseudosexual behavior of *cyanogaster* would seem to be a relatively new invention. Why was it selected for? Again the factor of crowding may have been determining. Other members of the genus *Coracias,* even though they are perchers and pouncers, probably have had a great deal of room to move around in for a very long time. Species of *Eurystomus* are aerial predators, the analogues of giant swifts, with almost unlimited space in which to maneuver. Only the ancestors of the blue-bellied roller must have occurred in clusters around clearings in the forests of West Africa a few thousand years ago. Apparently they had to take special precautions against the dangers of overcrowding. It is presumably, therefore, that their defensive threat behavior became particularly elaborate.

The fundamental point I am suggesting is that both some blue-bellied rollers and some pied kingfishers have developed distinctive patterns of sexual origin to control hostility. These patterns are not used in precisely the same ways in the two species. The "homosexual" reactions of male pied kingfishers seem to be designed primarily to reinforce friendly relations among the individuals involved, and thereby to reduce the probabilities of fighting among them. In effect, the behavior accommodates the performers. The reversed heterosexual patterns of blue-bellied rollers may also reinforce friendly relations among the individuals directly or immediately involved, but they seem to be designed primarily to control hostility between the performers and their nonperforming rivals.

There is a difference of emphasis.

Hammerkops

Multiple reverse mountings are common in an African bird, the hammerkop, *Scopus umbretta*. A member of the order Ciconiiformes, it is often put in a separate taxonomic family of its own, but it resembles storks in appearance and many aspects of its way of life. It is a long-legged, large-billed predator, mostly upon aquatic organisms. The general behavior and natural history of the species are discussed, with many references, in H. L. Brown et al. (1982).

Why so much reverse mounting? To what advantage? Why should hammerkops be so peculiarly convergent on blue-bellied rollers? The powerful bill must, of course, be managed carefully. Hammerkops are not particularly crowded. Yet there is little doubt that their reversals really are hostile. They are performed by members of pairs, but seemingly only when there are intruders and/or rivals in the vicinity.

Other factors may be involved. In particular, individuals build very large and spectacular nests with great expenditures of time and energy. Why they should make such efforts is still unclear. The nests are far too large to be merely receptacles for eggs and young. They do not seem to function as insulation chambers (R. T. Wilson 1992). They may provide some protection from outside attacks. They also attract other commensal species. They are sometimes used as observation posts, refuges, or nest sites by barn owls, falcons, Egyptian geese, small mammals, monitor lizards, snakes. Some of these "guests" may serve as sentinels, defenders, and/or decoys. The whole arrangement is a preview of a complex interspecific society. A nest can be the stationary base or focus point of a flock that is very mixed indeed.

Rather simpler societies of nest associates have been described for many other birds. Possible functions usually are fairly obvious. Thus, for instance, Lindell (1996) suggests that associates, other small passerines, help plain-fronted thornbirds *(Phacellodomus rufifrons)* in Venezuela by nest-guarding and enhanced mobbing of predators.

For whatever reasons, hammerkops seem to consider their immense constructions to be valuable resources. Certainly they must be expensive to build. Such resources should be protected, by somewhat extraordinary means if necessary or desirable. Blue-bellied rollers and hammerkops have evolved similar protective devices to cope with problems that appear to be rather different at an immediate level.

Woodpeckers (Picidae)

Other examples of reverse mountings may be relevant. Suggestive notes on some woodpeckers in the published literature are listed in Short (1982).

Reverse mountings are known or suspected to occur in the acorn woodpecker, *Melanerpes formicivorus* (MacRoberts and MacRoberts 1976), the red-bellied woodpecker, *M. carolinus* (Kilham 1961), the black-cheeked woodpecker, *M. pucherani* (Eisenmann, cited in Short 1982), and the lesser flame-bodied woodpecker, *Dinopium benghalense* (Neelakantan 1962). "False" or "fake" copulations, apparently without reversals, have been performed by red-cockaded woodpeckers, *Picoides borealis,* and hairy woodpeckers, *P. villosus* (Kilham 1966).

COMMENT: INCREASING RISKS

These bits and pieces of behavior are consistent. They point to a conclusion. The stronger the competition, the greater the prizes at risk, the more numerous the incentives and opportunities for attack, then the more nearly essential are the strongest possible controls. Sexually derived patterns are often the best available.

4
GREGARIOUSNESS

The patterns cited in chapter 1 seem to be designed to deter overt fighting, directly or indirectly. But many are themselves partly or wholly hostile in origin or motivation. This is not really a paradox. The aggressive components of these patterns are of low intensity and/or deflected in some way. They are supposed to be relatively innocuous *substitutes* for the really dangerous forms of attack.

Of course, social life is not all hostility. There are sexual patterns and various (other) familial interactions (sometimes extending beyond real biological families). In addition, there are many responses that seem to be partly or wholly nonhostile in causation, without being directly or exclusively tied to such activities as courtship, copulation, or care of young at the moment of performance (whatever their phylogenetic origin). These responses, when and if considered as a group, might appear to be categorized or identified negatively, by what they are not. But they are very common, and they are important factors in the establishment and maintenance of many social relations. There may be some partly distinct motivation involved. For want of better terms available in the literature, I shall refer to much of this behavior as "friendly" or "gregarious."

APPROACHING AND FOLLOWING

The simplest pattern of a possibly friendly nature is doing nothing in a social situation. This can be a signal at times. It certainly is economical (but also see the legend to fig. 12).

Fig. 12. Facial expressions accompanying high-intensity, largely aggressive vocalizations in the spider monkey, *Ateles fusciceps* (top), and the adult male howler monkey, *Alouatta villosa* (Moynihan 1976, 56, fig. 14).

Then there are movements. Most mammals and birds—all but the most nearly solitary types—approach, join, and follow other individuals frequently. In many cases, this does seem to be a positive expression of friendliness. I shall use the terms *gregarious* and *gregariousness* for such behavior. Other workers have referred to the same or similar patterns as *distance-decreasing* or *affiliative* (e.g., de Waal and other students of primates).

Joining and following movements can be simple or complex in physical form. Their costs must depend upon the speed and vigor with which they are expressed. In the "right" circumstances, they may discourage aggression by covering or suppressing provocative stimuli and/or by encouraging some relaxation of tensions or changes of moods.

Bonds of gregariousness can be extended to produce large groups, or even mixed groups of individuals of different species. Some of the possible developments are noted in this chapter along with more comments on probable evolutionary histories and adaptive significances.

Friendly or gregarious movements, approaching and following, can be developed to considerable, even spectacular, effects.

IMMEDIATE BENEFITS

These various patterns may result in a high degree of cooperation (Leigh, in press; and other authors). What is the nature of the cooperation? There is little or no evidence that the individuals involved are benevolent to one another. They do not need to be so. Insofar as one individual helps another, it seems to be doing so to help itself, in the short and/or long term. Some students of animal behavior have had trouble with this concept. They seem to believe that interindividual competition should or must, somehow, lead to conflict. Thus, for instance, in a very recent paper, in a very reputable journal: "In this view of the living world, conflict seems very natural and cooperation appears as a phenomenon that usually requires subtle explanation" (Hammerstein and Hoekstra 1995). And yet cooperation actually exists. It is, in fact, common. It must have been selected for.

Biologists who have observed herds or bands or flocks (or prides or gaggles) of animals in the field have suggested that gregariousness, in any given case, can *and does* provide any one or all of a variety of immediate practical benefits, for the individual and often for the group. The advantages most often suggested have to do with food and with protection from predators. Many eyes and ears may detect predators easily. Individuals may warn companions. Individuals may also help one another to beat off or to confuse predators. Similarly, they may cooperate to locate and catch prey, to flush insects, to find flowering or fruiting trees—to discover new and greener pastures. Gatherings may function as information centers (see discussions in Bayer 1982; Lefebvre et al. 1996; Richner and Heeb 1996). Individuals may assist one another in breeding and raising young. Harcourt (1987) discusses benefits to helpers in both birds and primates. He could have found many more examples (also see Dunbar 1995b).

ORIGINS

It may be useful to consider the possible or probable origins of joining, following, and other friendly patterns. They probably are not the same in all cases. Some cases are interesting in themselves, although perhaps tangential to the main argument.

For instance, many marine teleost fishes lay eggs and have no parental care. The young hatch independently and disperse and scatter immediately, often as plankton, in the water column. They may remain dispersed for days or weeks. Sooner or later, however, if they survive, they attempt to move (return) to suitable environments to complete their juvenile and adult lives. As they do so, they are likely to encounter other individuals of their own species. Individuals may approach one another and clump together to form schools. Since these individuals never saw their parents (or recognized them as such), their approaches to companions must be almost entirely innate (see Thresher 1984; Barlow 1985; Pitcher 1986).

Coleoid cephalopods are comparable. None of the species have parental care, at least after the eggs have hatched. Moreover, with very few exceptions (some abnormal), all the species are semelparous (i.e., adults die more or less immediately after reproducing once). The young appear to scatter. In some cases, they also reassemble in schools (references in Moynihan 1985).

Some other, distantly related, invertebrates such as some polychaete (jointed) worms also assemble by "choice" in appropriate circumstances (comments in Toonan and Pawlik 1994).

All this is good evidence, perhaps the best available, for the existence of something like an independent gregarious "drive."

Schools of the Caribbean reef squid, *Sepioteuthis sepioidea,* seem to be socially homogeneous. There are few or no indications of any internal differentiation of social roles, hierarchies or alliances, before the onset of courtship. Perhaps not coincidentally, there are few signs of overt hostility within schools. (Individuals show many hostile color patterns, but most of these seem to be provoked by the appearance or approach of potential predators.) Thus it seems unlikely that the gregariousness of the squids is much concerned with the control of intraspecific aggression. It must help in defense against predators: early warning, scattering effects, etc. It may also allow "monitoring" of companions and potential competitors (fig. 13). One individual is unlikely to be able to monopolize a valuable resource of any size, for example, a school of prey or a patch of turtle grass, without sharing the loot with some neighbors. This social monitoring is control of competition of a rather distinctive kind (also see Moynihan 1978).

Fig. 13. Schools of the Caribbean reef squid *(Sepioteuthis sepioidea)* showing various color patterns for defense against predators and/or monitoring companions. From Moynihan and Rodaniche 1977, 298, fig. 7.

"HIGHER VERTEBRATES"

Birds and mammals are as different from squid and some fishes in the ontogeny of their social patterns as in other respects. All mammals do have parental behavior. So do almost all birds, megapodes being the only (nearly) complete exception. Members of both classes have repeatedly evolved complex social structures of a more or less friendly nature, various kinds of gregariousness in the broad sense of the term as used here. Again the literature is enormous. Fortunately, we have good surveys of cooperative breeding by birds (Vehrencamp 1978, 1979; Emlen and Vehrencamp 1983; Stacey and Koenig 1990) and a recent discussion of affiliative behavior in nonhuman primates (a whole issue of the journal *Behaviour,* vol. 130, pts. 3–4 [1994], was dedicated to this topic, with papers by Hill and van Hoof, Strier, Mitchell, Boinski, Furuichi and Ihobe, te Boekhorst and Hogeweg, Starin, Silk, Hill, and van Hoof and van Schaik). Some monographic studies of animals as diverse as burrowing rodents and coraciiform birds also give particularly useful lists of references (e.g., Hoogland 1995; Fry 1984; Fry et al. 1992; Kemp 1995).

It will be argued here that many of the joining and following patterns of many gregarious birds and mammals are, in some sense (at least historically and/or ontogenetically), extensions or extrapolations of "familial" reactions. Individuals may react to others as if they were parents, offspring, siblings, or cousins. The scope is wide and various. Extended families and/or the derivatives thereof differ in size and structure according to composition and circumstances. Again, detailed examples—provided below—are crucial in that they provide insight into subtle expression of gregariousness characterizing the social behaviors of various species of birds and mammals.

After tree sloths and ground squirrels, it would be nice to consider the social behavior of ground sloths and tree squirrels. Unfortunately, the necessary data are either unavailable or unappealing.

Corvids

Some associations are fairly obviously derived from a real family nexus. Consider some of the jays and crows, well-known birds of the family Corvidae, order Passeriformes.

The behavior of the Florida scrub jay, *Aphelocoma c. coerulea,* has been described by Woolfenden and Fitzpatrick (1984). In this form (perhaps a full species), most young adults remain in their natal territories through their first breeding season (the year after hatching). The old and young individuals tolerate one another easily. The young act as "helpers at the nest" for their parents. This is real cooperative breeding. It is only after an interval, perhaps prolonged for several years, that some young eventually move out to pair and breed in other areas.

Another *Aphelocoma,* the Mexican jay, *ultramarina,* has a more complex system. It occurs in territorial flocks of considerable size and structural diversity. Several generations breed within a single territory at the same time (Brown and Brown 1981). This is communal as well as cooperative breeding.

The Corvidae are, in fact, something of a treasure trove for the student of gregariousness. There are interesting social arrangements in other New World jays such as the pinyon jay, *Gymnorhinus cyanocephalus,* and species of the tropical genera *Cyanolyca* and *Cyanocorax* (see, for instance, Hardy 1961; Kirkpatrick and Woolfenden 1986; J. L. Brown 1974, 1978; and Marzluff and Balda 1988). Many crows often occur in flocks or dense aggregations. Some even nest communally, in colonies. Sharing by ravens has already been mentioned. Perhaps the climax of one line of evolution is the Indian house crow, *Corvus splendens,* which is not only gregarious itself, but also has become "a confirmed commensal of man, almost an element of his social system" (Ali and Ripley 1970). As far as I know, none of the true crows has helpers at the nest. Emlen and Vehrencamp probably are correct in suggesting that cooperative breeding is correlated with shortage of territory openings or other spatial and social opportunities. There is a survey of all corvids in Goodwin (1976). The best accounts of the behavior of a single species of crow are still those of Lorenz (1931, 1952) on the jackdaw, *Corvus monedula.*

Piciformes and Coraciiformes

Other groups of birds vary in much the same ways as corvids. Some species are also cooperative and sometimes communal in their nesting habits. This is true, for instance, of populations of the acorn woodpecker, *Melanerpes formicivorus* (MacRoberts and MacRoberts 1976; Koenig 1981; Koenig and Mumme 1987; Bennun and Read 1988), as well as a rather surprising variety of coraciiforms, mostly bee-eaters, woodhoopoes, and hornbills. It may be supposed that the habit of cooperative breeding has been evolved independently on several occasions in the evolution of different lineages of Coraciiformes.

Bee-eaters are comparatively small in size. They are almost exclusively insectivorous (a few individuals have been seen to take small fishes from near the surface of the water on a few occasions). They get their insects in various ways. Some of them are sit-and-wait types, but most of them usually catch their prey by aerial forays rather than by pouncing. Some of the most aerial forms fly back and forth, "hawking" for insects, for long periods of time. These species often occur in flocks and feed gregariously. Some of them also nest in colonies. Fry (1984), who recognizes twenty-four species in the taxonomic family Meropidae, says that fourteen of them sometimes or usually breed in loose aggregations, while another four of them breed in dense or tight colonies of tens, hundreds, or even thousands of nests.

The breeding behavior of one of the highly gregarious species, the white-throated bee-eater, *Merops bullockoides,* has been studied at length and in detail by Emlen and his collaborators (Emlen 1981, 1982a, 1982b, 1990; Emlen and Demong 1980, 1984; Emlen and Wrege 1991, 1992; Wrege and Emlen 1994). Social relations are complex. There are definite, structured subgroups within the colonies. These have been called "clans." Adult individuals shift from helper to breeder roles, and then back again, often repeatedly. Reciprocal exchanges of helping are fairly common: an elaboration of affiliations.

Fry notes that there is still a great deal of squabbling and disputing in densely packed colonies. This would suggest that the birds have been able to get together because their gregarious tendencies have gone up, not so much because their aggressiveness has gone down.

Woodhoopoes are also comparatively small and insectivorous or omnivorous, but they get most of their food by scrambling, probing, and gleaning in trees and bushes. They do not usually pounce; neither do they wait. Still, the breeding behavior of one species, the green woodhoopoe, *Phoeniculus purpureus* (sic!), is reminiscent of the white-throated bee-eater: colonial with much helping and reciprocity (Ligon and Ligon 1978a, 1978b, 1981, 1983, 1990; Ligon and Stacey 1989; Du Plessis 1991, 1993).

It is not really known why some bee-eaters and the green woodhoopoe have carried communal life to such a pitch. Fry suggests that the prey of some bee-eaters are so widely and irregularly distributed that the birds are unable, or have no incentive, to set up feeding territories. Without such territories, the advantages of gregariousness (e.g., easier detection of predators) may appear to be particularly attractive or rewarding. Apart from this, one can only speculate, again following the suggestion of Emlen and Vehrencamp that the birds have difficulty in finding nest sites. The woodhoopoes must look for holes and cracks in trees. The bee-eaters need bare cliffs or earthen banks in which to excavate holes. Soils suitable for digging may not be easily accessible everywhere.

The gregariousness of some species is all the more remarkable because there are other forms of *Merops* and *Phoeniculus* that definitely do not breed in groups, or even assemble in flocks (Fry 1984; Fry et al. 1988).

Hornbills are supposed to be related phylogenetically to woodhoopoes, but they are much more diverse than woodhoopoes or bee-eaters. They range in size from medium to gigantic (by coraciiform standards). Some forms are frugivorous-omnivorous. Others tend to be insectivorous or even partly carnivorous. Different species are variously distributed in forests and/or open country. The group as a whole has been reviewed by Kemp (1995). His generic and specific nomenclature is followed here.

Kemp recognizes fifty-four species of hornbills. Many, perhaps all, of these are gregarious, at least occasionally. There is also cooperative breeding. Kemp says that breeding groups of a pair with helpers have been recorded for eight species and are suspected to occur in at least ten others. The helpers are all males in some cases.

Aspects of the biology of Bornean hornbills are described in Leighton (1982, 1986) and Leighton and Leighton (1983). Nests are more or less scattered; there are no colonies. Nevertheless, helpers at the nest are tolerated or encouraged by adults in two species, *Anorrhinus galeritus* and *Aceros (Berenicornis) comatus.* Individuals of a third species, *Buceros rhinoceros,* can be facultative helpers in emergencies. Leighton suggests that most of the suitable territories are preoccupied, at any given time, by breeding pairs of great longevity.

Carnivora

Some mammals may be gregarious in much the same ways as some birds. This seems to be true of the carnivores cited and discussed by Kruuk (1989). In this group, there are many species whose young from previous litters help in rearing their mothers' subsequent offspring during the following years. Examples include jackals and wolves, foxes, lions, and hyenas. These animals often take relatively large prey. Attendants may be really helpful in bringing chunks of food to the infants of their siblings.

It should be noted, however, that the young of previous litters of the Eurasian badger *(Meles meles),* at least in Britain, may also stay with their parents for appreciable periods of time after reaching maturity without, apparently, helping at all.

Again, it seems that the older young are waiting for suitable territories to become available before moving out on their own.

Primates

There may also be examples of parallel and convergent evolution of gregariousness among primates.

The lemurs of Madagascar (perhaps five taxonomic families: Cheirogaleidae, Lepilemuridae, Lemuridae, Indriidae, and Daubentoniidae) show a full spread of social arrangements, from nearly solitary types through family groups and small multimale groups to large, multimale groups (Tattersall 1982; also see Jolly 1961; Petter et al. 1977; Richard 1978; Charles-Dominique et al. 1980; Kappeler and Ganz-Horn 1993). They are so diverse that they reveal few general rules. Tattersall can be quoted directly: "The social groupings displayed by the Malagasy primates vary widely, not simply between but in some cases also within genera and even species. Neither phylogenetic relationships nor ecology seems by itself to be an efficient predictor of the size or the organization of social units, although as a rule rain forest lemurs tend to live in smaller groups than do related forms inhabiting drier, more seasonal areas." Dominance hierarchies can be detected within some social groups. They seem to be relatively straightforward, linear rather than intricate.

The monkeys and apes, simians and anthropoids, suborder Haplorhini, are also extremely diverse (even without man), but not in quite the same ways as the lemurs. At least the percentages of social types are different. None of the animals is nearly solitary, with the possible intermittent exception of adult male orangutans, *Pongo pygmaeus* (MacKinnon 1974). There are a number of species, at almost opposite extremes of the haplorhine radiation, that are usually found in pairs or in small family groups. Among them are some ceboids; the night monkey, *Aotus;* titi monkeys, *Callicebus* spp.; and Goeldi's tamarin, *Callimico goeldii;* other tamarins, *Saguinus* spp.; marmosets, *Callithrix* spp. (Moynihan 1966, 1976; Mason 1966; J. G. Robinson 1977; Kinzey 1981; Menzel 1993; Dunbar 1995a); and all the "lesser" apes, the gibbons and siamang of the genus *Hylobates* (references in Chivers 1980).

The rest of the monkeys and apes are more or less highly gregarious, usually occurring in bands or troops. The sizes of the groups vary widely according to species and circumstances. Groups also show differences in density, cohesion versus dispersion. There are records of age cohorts, dominance hierarchies, alliances and partnerships, phatries, troops of "bachelor" males (apparently unmated at the time), harems, and multimale (and female) assemblages. Again, the published literature is very large. I can make no attempt to cover it. Some publications, well-known monographs, surveys, and a few relevant papers can only be listed: Carpenter (1934), Schaller (1963), Imanishi and Altmann (1965), Kummer (1968), van Lawick-Goodall and van Lawick (1971), Rowell (1971, 1972), Angst (1974), Seyfarth (1976), Vogel (1976), Blaffer Hrdy (1977), McKenna (1978), Hinde (1983), Smuts (1985), Cords (1988), Gautier-Hion et al. (1988), York and Rowell (1988), Cheney and Seyfarth (1989), de Waal (1989), van Schaik (1989), Harcourt (1992), Peres (1992), Mason and Mendoza (1993), Boinski and Mitchell (1994), and Watts (1995a, 1995b).

A point of some general interest is raised by Rowell (1988) in comparing the systems of communication of different kinds of African monkeys. She says: "There seem to be two parallel ways of coordinating social groups: the first is by the exchange of overt, specialized signals (gestures, noises); the second is by each individual monitoring others' movements and to what they are paying attention, and adjusting its own position accordingly." She suggests that the guenons, *Cercopithecus* spp., and close relatives exemplify the first strategy, while the papionines, baboons, macaques, and mangabeys prefer the second. Doubtless, similar differences in temperament could be found in a wide range of mammals and birds.

Many or most primates probably derive all the benefits that are usual to the habit of gregariousness. But why in the world do they bother to do so? There is a lot of maneuvering and manipulation involved. It is not absolutely necessary.

Tree Sloths

There might be supposed to be alternatives to gregariousness. One possible alternate comes to mind, exemplified by a group of edentates. The sloths of the order Xenarthra provide a standard of comparison and perhaps an object lesson. They are arboreal vegetarians of lowland and hill forests of the New World tropics. More precisely, they are folivores, leaf-eaters. The two-toed sloths, *Choloepus,* eat leaves and a few fruits. The three-toed sloths, *Bradypus,* eat only leaves.

There are, of course, folivorous primates, for example, the howler monkeys *(Alouatta)* of the New World, the langurs *(Presbytis)* of southern Asia, and the guerezas *(Colobus)* of Africa. They are often very abundant, and their social behavior is quite typically monkeylike in most respects, not slothful, as a low-energy diet of leaves might be thought to require.

Not so the behavior of sloths. These animals are as nearly solitary as possible. Males and females associate only briefly. Females carry the young for only a few months. There would seem to be some sort of mutual avoidance among individuals, although probably not overt territorial defense. The two-toed sloth is nocturnal. The three-toed sloth is also active by day. Both types have relatively low body temperatures. They might be described as partly ectothermic. Individuals of both types move very slowly (hence the name). All this is very low key. But the animals are extremely successful. Their food is always at hand. The climate is favorable on the whole, although not absolutely benign. They are protected from predators by their habits and inconspicuousness. (Actually, the three-toed sloth hardly attempts to defend itself if and when it is seized by an unnatural predator, a curious biologist.)

The net result is that the animals are not only numerous as individuals, but they also make a significant contribution to the local biomass. According to Montgomery and Sunquist (1975, 1978), in at least some areas the biomass of three-toed sloths alone is greater than that of howler monkeys.

Obviously, there are alternative routes to the same "goal." This may be another general rule of social life. Thus, large populations can be built up by animals often with similar feeding habits, and based on either complex social groups (at one extreme) or a virtually solitary lifestyle (at the other). The selection of one alternative, rather than another, probably is not entirely free. It is unlikely to be a purely behavioral choice on the spur of the moment, or even the product of chance mutations alone. There may be other factors in play: constraints of background or objective "predispositions." Consider the tree sloths and folivorous primates again. The ancestors of the modern primates may still have been small insectivores when they first took to life in the trees. The immediate ancestors of tree sloths, on the other hand, may have already become larger and herbivorous before they became arboreal. Differences such as these, with their inevitable consequences, for example, different degrees of vulnerability to predation, might be sufficient to explain, at least in part, why the social relations of the two competitive types were subject to different selection pressures.

Historical scenarios can be plausible. They probably should not, however, be elaborated or carried too far (see the comments on the dangers of historical reconstructions of behavior below).

Ground Squirrels

The rodents, usually lumped together under the general vernacular name of "ground squirrels"—species of *Spermophila* and *Cynomys,* and the related marmots, *Marmota*—all members of the taxonomic family Sciuridae, share many behavioral as well as anatomical features. As a taxonomic group, they are much less diverse than primates; but they also have evoked considerable interest. There are published observations and studies of at least nine forms of marmots and seventeen species or well-marked subspecies of ground squirrels. Intensive studies of one ground squirrel, the black-tailed prairie dog, *Cynomys ludovicianus,* are summarized in Hoogland (1995). The species is herbivorous, diurnal, burrowing, and gregarious. Individuals live in permanent colonies (popularly called "towns" or "villages"). Colonies can be very large, with thousands of residents, in undisturbed conditions. The residents dig many burrows, often very extensive. Colonies and their tunnel systems may spread for kilometers in all directions. Individuals use their burrows for resting and sleeping, for protection from both predators and extremes of weather, for the performance of some social actions, including most copulations, and for the rearing of young. But feeding occurs almost exclusively above ground.

Colonies are divided into wards and smaller subgroups: "coteries." The latter may be roughly equivalent to the clans recognized in some other animals. In any case, among these black-tailed prairie dogs (i.e., ground squirrels, *Cynomys ludovicianus*), coteries are polygynous families. They consist of one or (not infrequently) two adult males, two or three adult females, and one or two juvenile offspring. The offspring may be of either sex. All the adults are capable of breeding. Apparently they may all do so in the same season. A large colony has many breeders.

Both breeding and nonbreeding individuals can be very aggressive, occasionally to other members of the same colony, more frequently to members of other colonies. Their aggressiveness is mediated by vocal, visual, tactile, and (presumably) olfactory signals (also see W. J. Smith et al. 1976, 1977).

The black-tailed prairie dog would seem to be more highly specialized, socially, than the other ground squirrels and marmots that have been studied. In this connection, Hoogland proposes a rough classification, three different "levels" of group living. Again, proceeding from simple to less simple: (1) mere temporary clumps of individuals of the same or different species; (2) groups of conspecifics that live in the same area during parts of the year; (3) groups of conspecifics that live together throughout the year, from generation to generation.

CORRELATES OF EUSOCIALITY IN MAMMALS

By similar criteria, the next "higher" step of group living is eusociality. Quite "ordinary" social signals can be surprisingly effective even in bizarre social organizations. A few species of mammals, only distantly related to one another phylogenetically, have carried gregariousness to a stage or level beyond that represented by the black-tailed prairie dog. Not only do these species occur in persistent and highly integrated groups, they have also, like some insects, for example, termites and ants and many wasps and bees, become "eusocial." Only one female, the alpha female or queen, breeds successfully in any given group or colony, under natural conditions, at any given time. The best known cases are the dwarf mongooses described by Rasa (1985, 1994) and the East African naked mole rat, *Heterocephalus glaber,* which has been studied by P. W. Sherman and his collaborators (see the publications of 1991 and 1992). The Zambian naked mole rat, *Cryptomys* sp., also seems to be eusocial (Burda, pers. comm.). (Mole rats belong to the taxonomic family Bathyergidae. All the members of the taxon are specialized in many respects, but they are not all eusocial. *Spalax,* for instance, definitely is not.)

The animals are more or less difficult to observe. Mole rats are subterranean burrowers. They virtually never come to the surface. Dwarf mongooses are often seen on the ground, more or less in the open; but they also spend much of their time, at night and in the heat of the day, resting and sleeping in the ventilation chambers and galleries of large termitaria.

The cramped quarters of the animals may have had something to do with the development of their eusociality; but the causal connection, if any, has not, to my knowledge, been traced or demonstrated very convincingly.

There is some division of labor. A group of dwarf mongooses includes, in addition to the queen, an alpha male or consort to provide a degree of leadership as well as a supply of sperm, plus a variable (not very large) number of other adults and young of both sexes. Juveniles and nonreproductive adults may act as sentinels, scouts, nursemaids, and defenders. In colonies of *Heterocephalus,* which may include up to eighty individuals, the nonreproductives come in two sizes. Small individuals dig burrows and search for food. Most large individuals remain with the queen, perhaps helping to take care of the pups, but some particularly fat males tend to disperse away from the colonies (O'Ridin et al. 1996).

All this is very reminiscent of the castes of eusocial insects.

Eusociality is a highly derived condition. It is interesting, therefore, that it occurs with—is at least partly maintained by—social signals that are themselves quite "ordinary" in quality. The dwarf mongooses, for instance, have a repertory that seems to be typically carnivorelike, if perhaps slightly on the elaborate side. They do a lot of olfactory signaling, allogrooming, and playing. Their visual signals include the usual postures, movements, and pilo-erection patterns. They utter many different sounds (even above ground, individuals of a group are often out of sight of one another in thick scrub).

The communication system of *Heterocephalus* may be simpler. But queens use aggressive butting to keep other members of the colony hard at work (Reeve 1992).

Whatever may have been the selection pressures that promoted the development of eusociality, it would appear to be very unlikely that the process was initiated, pulled forward, or pushed from behind by any intrinsic features or peculiarities of the signal systems acting as anything like an independent motive force.

Signals, like other tactics, are available and indispensable. They might, perhaps, be compared with the morphemes and phonemes of human language, or with the tools and materials of human construction. No sentences can be formed, nor buildings erected, without them. But the same elements can be used to different ends. Traffic signs or deathless prose. The Lincoln Memorial or the Washington Monument. The solitary life of the adult male orangutan or the elaborate societies of some bees and bee-eaters. Signals, by themselves, provide no direction or opportunities for social evolution.

DIGRESSION: A DISTINCT SYNDROME?

Is there life under ground? Well, yes, of course. But for many mammals, and perhaps especially for rodents, the adoption of more or less subterranean habits is combined with particular social patterns and relations: the development of highly gregarious or colonial groups, a proliferation of infanticides within such groups, and, in some cases, the establishment of eusociality. This seems to have occurred independently in several different phylogenetic lineages (even among rodents, the ground squirrels and mole rats are only very distantly related to one another).

Some living species may represent intermediate stages of this general trend, the evolution of this cluster of characters. Thus, for instance, there are cases of infanticide that fall between the neat categories listed above. According to Hoogland, female black-tailed prairie dogs frequently kill the offspring of close kin as well as those of unrelated or distantly related individuals. Perhaps the species is on the way to attaining eusociality.

Once attained, eusociality may be maintained by various means. Infanticide seems to be used by dwarf mongooses. In other species, individuals may be kept in nonreproductive condition by nonviolent behavioral or physiological controls.

Why should there be this general trend among burrowing animals? (Note that the excavation or arrangement of passages and chambers in termitaria, as practiced by dwarf mongooses, probably is functionally equivalent to burrowing into the ground directly.) Again, the answer may be related to spacing or, more precisely, crowding.

Any system of tunnels or burrows, however extensive or scattered (termitaria), must be confining or restrictive. Even when the owners come into the open frequently, they cannot go too far away from their repairs or refuges. When they do not, as is supposed to be the case with mole rats, the effect may be claustrophobic as well as reassuring.

It seems likely that crowding, in groups or colonies of whatever sort, is correlated with a scarcity or shortage of resources. The correlation may well be causal; but the connection has not been traced in detail. Either the groups or colonies are established because necessary resources occur in patches, and/or the concentration of individuals in a group or colony tends to exhaust the surrounding resources in patchlike patterns.

Crowding is frequently combined with—it may even be a cause of—an intensification or exaggeration, hypertrophy, of either aggression or the controls of aggression or both. This is not very surprising. But the elaboration takes different forms in different cases. There is infanticide and eusociality in some burrowing mammals. There is cooperative and communal breeding in some birds. There also are peculiar displays. Some of these are discussed in the next section.

KINSHIP GROUPS, AND SOME RELATED AND CONVERGENT PHENOMENA

Observers and analysts have paid increased attention to the patterns and responses of related animals, their interactions among themselves, almost from the moment of appearance of Hamilton's two earlier, seminal and very influential papers on the genetical evolution of social behavior (1964a, 1964b).

It has been confirmed that many social groups of birds and mammals are based upon—or at least include—some individuals that are closely related to one another by descent. For these animals, the subject is usually discussed in terms of parents, offspring, and siblings. Such individuals often help one another. Theory predicts that they should do so because they have so many alleles in common by descent. There also are practical considerations. Close relatives are among the individuals most likely to be around in the neighborhood when help is needed. They also are likely to be familiar with one another, to have established relatively peaceful social bonds among themselves.

These bonds, by themselves alone, are not always appropriate or sufficient for the most effective social structures to be organized.

Thus, for instance, there often are precautions against "incest." In many birds, presumably leaving aside some of the communal or cooperatively breeding types, the original male and female founders of a new family or lineage probably are not very closely related to one another. This could be one (among several) of the reasons why so many male birds sing loudly: perhaps they are trying to attract females from afar.

There may be other mechanisms. Consider the hyenas once again, especially the brown hyenas of the southern Kalahari studied by Mills (1989). Some of these animals live in groups or clans including both males and females. There also are nomadic males. They range widely and do not belong to any clan. They are not, however, useless. Mills says that it is the nomadic males, and not the clan males, that mate with the group-living females. This system may favor outbreeding, but it can hardly fail to increase some kinds of social tensions within clans.

Incest avoidance is supposed to be the proximate cause or mechanism of eusociality in the Zambian naked mole rat. According to Burda, the avoidance is based upon individual (not only family-member) recognition. Females, other than the queen, do not mate, in part because all males are relatives (i.e., fathers or brothers).

Apart from inbreeding restrictions, both cooperation among kin in the short run and kin selection in the long run should be free and clear to continue indefinitely.

Actually, they have run very widely. There is no doubt that kin selection must be, and must have been, significant in the evolution of many kinds of social organizations, even including those that are not openly gregarious or cooperative. It does not, however, explain everything. There has been a tendency to overemphasize or overestimate its importance in particular cases.

Thus, for instance, Agrell (1995) went to the trouble of checking out the possible role of kin selection in producing changes in the social organization of female field voles *(Microtus agrestis)* between seasons or over a single season. Kin selection in a mammal should not really have been expected to be detectable in such a short time. And, in fact, Agrell comes to the quite sensible conclusion that food availability and predation probably were more relevant to his animals in the circumstances.

Rather similarly, Hare and Murie (1996) found that kinship has little to do, directly, with social discrimination in groups of juvenile Columbian ground squirrels (*Spermophilus columbianus*). Reciprocal altruism and "dear enemy" relations seem to be sufficient to explain most of the groupings observed among these animals.

A few extra comments may be useful.

First of all, it is obvious that close kin are often too scarce, not numerous enough, to fill all the social roles available or even necessary for the survival of a group. Thus, one frequently reads of strangers, interlopers, immigrants, nonkin, and so on in groups as diverse as clans of hyenas, harems of langurs (references above), clusters of cooperatively breeding birds (e.g., Ligon 1983; as well as earlier references), and roosting assemblies of black vultures (Parker et al. 1995). The evidence, which is fairly good, does indicate that the individuals involved are not parents, offspring, or siblings. It does not, however, exclude the possibility that many of them are cousins, who must also, of course, share some of the same alleles by descent as their associates. It seems likely that, in appropriate circumstances, many social groups include remote (or "nondescendant") kin in addition to close kin and nonkin. Degrees of relatedness may be difficult to calculate during field studies, but they should be assumed to be various.

It might be noted, parenthetically, that close kinship does not always produce the same social results in different species, even when the species themselves are closely related to one another. Remember the bee-eaters and woodhoopoes. There are other examples. Troops of ateline monkeys, the spider monkeys *(Ateles),* woolly monkeys *(Lagothrix),* and woolly spider monkeys *(Brachyteles)* include "brotherhoods" of philopatric males, but relations within groups of male *Brachyteles* are much more egalitarian than relations within the corresponding groups of the other two types (Strier 1994).

There are ties among females as well as among males in many species of animals; but it is male bonds that have drawn most attention (perhaps because they tend to be conspicuous?). Brotherhoods certainly occur in birds as well as in mammals. Real brothers have been supposed to be linked in North American turkeys, *Meleagris gallopavo,* in southeastern Texas (Watts and Stokes 1971) and are known to be in Tasmanian "native hens," *Tribonyx mentorii* (Maynard Smith and Ridpath 1972).

In a rather different case, feral Reeves's pheasants *(Syrmaticus reevesi)* form unisexual groups, males with males, females with females, in parts of western Europe and North America (Moynihan 1995; also see Biadi and Mayot 1990; and Knoder and Baillie 1956). The members of such groups must be related, at least cousins, in actual fact here and now. All or most of the Reeves's pheasants surviving in the Western world are descendants of a few, perhaps only one, successful introduction of the species from China in the late nineteenth century (Beebe 1926; Delacour 1951). It might also be noted that accounts of Himalayan monals, *Lophophorus impeyanus,* suggest that male bonding is a recurrent phenomenon among pheasants (Beebe 1918–26; Baker 1930; Johnsgard 1986).

Perhaps so. But all I really saw in eight years of observation of Reeves's pheasants were adjacent individuals joining and following one another. Propinquity must be a common cause of, or encouragement to, the formation of groups. Depending upon species and situations, neighbors may or may not be relatives. It is claimed, however, that some exceedingly diverse animals, for example, birds, bees, tadpoles, as well as naked mole rats, can identify one another individually (Pfennig and Sherman 1995). The clues, probably usually olfactory or visual, must differ in different cases.

Of course, personal recognition can also trigger flight and dispersion.

Be that as it may, one thing is certain: there *are* complex groups that are largely composed of nonkin. The best examples are mixed groups of different species. Mixed flocks of birds have been known for a long time (descriptions and references in such papers as Buskirk 1976; Buskirk et al. 1972; Gradwohl and Greenberg 1980; Krebs 1973; Moynihan 1962b, 1978, 1979). There are also mixed groups of mammals, for example, among the more arboreal forest guenons (Gautier-Hion 1988), and even associations between birds and mammals.

Some mixtures of birds may appear to be casual, accidental, or adventitious at first sight—thus, for instance, groups of hornbills *(Tockus erythrorhynchus),* starlings, coucals, woodhoopoes, shrikes, and other birds in semiarid savannah regions of Sénégal. The appearance is misleading. The same species, and probably individuals, reassemble day after day. They are not the mere "temporary clumps" cited by Hoogland (1995). Other mixed flocks are not only highly organized but quite obviously so. In many flocks of neotropical birds, mostly passerines, different species consistently play different social roles, for example, active nuclear, passive nuclear, and regular attendant.

A good example of elaboration is provided by one kind of mixed flock of the cold humid zone of the Andes. Flocks of this type are composed of brush-finches such as *Atlapetes* spp.; bush-tanagers and bush-warblers, for example, spp. of *Hemispingus, Chlorospingus,* and *Basileuterus;* bright tanagers of such genera as *Anisognathus* and *Buthraupis;* honeycreepers, mostly conebills of the genus *Conirostrum;* whitestarts, spp. of *Myioborus;* and many other birds such as tyrant flycatchers (Tyrannidae), and woodcreepers and spinetails (Furnariidae).

These species also show geographic variation in social behavior. As in the Diglossa cluster, there are consistent average differences between the birds of the center and those of the extremities or outliers of the zone. Only the actual patterns involved are quite dissimilar in the two cases (see Moynihan 1962a; and also Poulsen 1996).

The feeding advantages obtained by memberships in mixed groups are sometimes conspicuous. Thus, for instance, the *T. erythrorhynchus* hornbills catch insects and (probably) small vertebrates flushed by their companions. Individuals of the related species, *T. camurus,* regular members of mixed-bird parties in the humid forests of West Africa (very different from savannahs), catch insects disturbed not only by their avian companions but also by tree squirrels and army ants (Bovet 1931).

Relations can be sophisticated. Two other species, *Tockus flavirostris* and *deckeni,* are known to associate with dwarf mongooses (Rasa 1983). They catch insects, especially locusts, flushed by the mammals. They do not leave much to chance. They positively do everything to encourage the mongooses to "produce." They have special "chivying" and "waking" patterns to ensure that the mongooses do not fall behind their (the hornbills') schedule. Even more remarkable, they give warnings of flying predators, supposedly including even those hawks and eagles that are dangerous only to the mongooses! This is real cooperation, but certainly not altruism, because everyone benefits.

A more general note on *Tockus* may be interesting. Individuals of the genus occur as pairs or as small nuclear families, sometimes several of them, when associated with individuals of other species. They do not, however, carry gregariousness any further during the breeding season. Unlike some other hornbills (references above), they do not have helpers at the nest. The best general account of various species of the genus is by Kemp (1976).

In some cases, resemblances of color seem to have become exaggerated as a special adaptation to facilitate associations of different species. This might be called "social mimicry" (Moynihan 1968; also see Wickler 1968). Among small birds, there are indications that individuals of different species are most likely to come together if they share certain conspicuous characteristics such as color, for example, black and yellow, blue and green.

Social mimicry must depend upon several factors. Thus, for instance, it is less common among coral reef fishes than among birds (Moynihan 1982b). This difference seems to be correlated with spatial parameters and individual distances, the frequencies with which individuals are brought face to face with one another suddenly.

Instant recognition of suitable species to join may be needed for several reasons. It can be facilitated in various ways (also see the discussion of sexual dimorphism in kingfishers in chapter 3).

It has already been suggested that many of the friendly or affiliative patterns of birds and mammals have been derived, in some sense (phylogenetic and/or ontogenetic), from patterns that were originally evolved to regulate and encourage family (possibly including sexual) relations. Where else could they have been derived from (assuming, of course, that they come from somewhere within the preexisting repertory of a species)? The alternative, origination de novo, seems to have occurred in the gregarious cephalopods and some fishes. But this latter course might well be less easy than some others. It would be almost entirely dependent upon genetic accidents or recombinations. On the other hand, the extension or extrapolation of family behavior to nonfamily members, even to individuals of other species, would be hardly more than a reorientation.

"Strangers" may be fully accepted, at least eventually, within groups. To my knowledge, there is little or no published evidence that newcomers are discriminated against after the initial period of insertion.

5
EVOLUTIONARY CHANGES IN MOTIVATION AND FUNCTIONS OF SIGNALS

All things change, including social relations and patterns of communication, hostile as well as nonhostile. They all have origins, histories, proximate and ultimate causes, immediate and eventual effects or consequences. Each of them also, of course, has several aspects or dimensions. These can change independently of one another, at different rates in different species and in different circumstances.

Here we are primarily, almost exclusively, concerned with changes over the long term, in evolutionary time. The argument is meant to be logical, couched in ordinary, nontechnical words. Mathematical or strictly genetical modeling is left to other works (a very recent and rather typical example of which is Tanaka 1996). Long-term changes are difficult to assess. They can only be estimated post hoc, by comparisons among the patterns of related or analogous living species.

We also are primarily concerned with changes in behavior patterns of what might be termed small to medium "size" or scope. We are not really concerned with the transformations of entire social systems as such, but rather with modifications of the particular signals and controls that make up the systems. Again, the method must be comparative.

In principle, it should be possible to draw distinctions among changes in form; changes in causation, that is, motivation; and changes in consequences or effects, that is, functions. Naturally this is easier to do in some cases than in others. It also is more interesting in some cases than in others. It is hardly necessary or useful for some of the most common patterns.

For example, retreat, avoidance, and exclusion are simple. They encode information; they are potential signals. But they are unritualized or only very slightly ritualized per se. Their forms, apart from orientation, may be supposed to have changed little and slowly over time. Their motivation probably has been equally conservative. Of course, their actual or practical effects must differ according to circumstances.

There are many other simple motions and movements, for example, various approaches, that are also transparent and straightforward.

Even some patterns that are not quite straight literally may be easy enough to analyze. Thus, for instance, the movements of redirection are quite ordinary in themselves, but they are deflected. The reason for the deflection usually is obvious when performed in natural contexts. If some performances have become stereotyped, slightly ritualized, it merely adds to their effectiveness as signals.

These humdrum behavioral patterns could be said to belong to a sphere that French historians call *évenementielle* (anecdotal narrative, the equivalent of diplomatic history). Other patterns are not so easily dismissed. Perhaps they should be called *éxistentielle* (the real stuff, life as it is lived).

Analysis of causation can become difficult when "additional" kinds of motivation are involved or activated simultaneously. The difficulty is progressive. In many cases, tendencies such as attack, escape, to be friendly, to form pair bonds, can be accommodated together—one might almost say reconciled with one another—at low to moderate intensities of activation. There is no physical impediment. Even diverse movements and intention movements can be combined easily enough as long as they are all slight. But the process of accommodation has its limits. Full-scale expressions of incompatible tendencies are difficult or impossible to combine in any useful manner. Something else has to be tried, sooner or later, as intensity increases.

This is where diversions are likely to appear.

Some of these diversions can be interpreted only partially. Thus, for instance, the sexual soliciting patterns of adult birds remain in historical limbo. It is at least evident that their proximal causes and functions cannot be the same as those of the related and similar-looking patterns of nestlings such as crouching and upturned gape to beg food. The adult versions have become appeasement and/or sexual stimuli. Still, the derivations are not entirely clear. We seem to know both the starting points and the current positions; but the courses of the paths between them remain obscure to us. More is known or suspected of other performances.

COMBINATIONS

Consider the elaborate and obviously compound choking displays of gulls. These performances include scraping and turning movements, obviously derived from nest-building and nest-arranging, in addition to downward pecking movements and a characteristic vocalization. The full display, as such, is confined to the breeding season, but the correlation is general rather than particular. It often occurs during hostile conflicts with neighbors and/or as a reaction to hostile intrusions. In these circumstances, the nesting-type movements are not, at the moment of performance, closely or directly tied to the shaping of real nests in preparation for the reception of real eggs. They must have a special significance at such times. In fact, they appear to have become hostile, distinctly unfriendly expressions of attack and escape tendencies. They seem to be used in threat, probably largely defensive in purpose. It is possible, however, that they are not, or not yet, *entirely* hostile.

A change from partly or primarily nonhostile to purely or largely hostile motivation is unmistakable in the display patterns of sexual derivation. The reverse mountings of some blue-bellied rollers during the nonbreeding season can no longer be strictly copulatory in causation. Like elements of choking, they have become hostile in some sense. In what sense, and to what extent, are the pertinent questions. They are not easy to answer.

It is evident that the evolutionary traffic between hostile and nonhostile signal patterns is two-way, perhaps even repeatedly back-and-forth. Thus, for instance, patterns of hostile origin can be incorporated into performances and sequences that are not primarily hostile anymore. This would seem to have happened during the evolution of some of the "courtship" displays of gulls and terns (Tinbergen 1960; Moynihan 1962a).

We are talking a lot about reverses, converses, obverses; but the switches to which these terms apply do seem to be real. Evolution is not a one-way street.

Some of the possible modifications are again revealed by the blue-bellied rollers. Individuals utter pumping notes during what appear to be normal, functional copulations. These notes are harsh. They sound, in tonal quality, as if they were derived from the rasps and rattles uttered as threat during hostile encounters between opponents or rivals. The sexual versions must reflect, in some way, an original hostile component in the relations between males and females. This may no longer be true. The pumping rhythm is distinctive in form, and it seems to be confined to copulations and pseudo-copulations. During true copulations, the sound may have come to express copulatory motivation, at least in part. It can hardly fail to function as a sexual signal in these circumstances. But then it was transferred, along with the other sexual patterns, to the hostile pseudocopulations. Thus, elements of this closely related cluster of vocalizations has gone from hostile to sexual, or partly sexual, to hostile once again. *Éxistentielle* indeed!

Diversions of this kind must tend to be problematical. The first appearance and early occurrences of a diverted pattern in its new social context could prove to be awkward, counterproductive, or even dangerous if anything should go wrong. The timing and circumstances have to be right, and the "handling" of the introduction or insertion should be cautious and careful. A certain amount of good luck may also be required.

POSSIBLE HISTORICAL SEQUENCES

The next question is obvious. How, in fact, are diversions likely to have originated? Two possible scenarios come to mind. As new displays in their new circumstances, they could either have appeared abruptly, as a sudden surprise, or gradually, by a slow slippage or reorientation in time (rather than space).

One could imagine that something like the pseudosexual behavior of the blue-bellied roller could have appeared "fully fledged," a transfer from real sex, as a displacement pattern in the midst of a hostile performance. It is well known that displacement patterns occur at moments of stress, and the hostile encounters of the species are stressful to a degree.

Any abrupt appearance of this sort would have effects. First of all, it would be startling. This, in itself, probably would interrupt, however briefly, the pursuit of active hostility. The intrusion might be repeated. Perhaps the first repetitions would also be startling or distracting. The effect must wear off with further repetitions. Eventually, probably sooner than later, both the performer and the observers of the performance would have to accept and adapt to the new behavioral situation. In the case of the blue-bellied rollers, the most pertinent and important observers, the territorial rivals, have come to "understand" the hostile messages encoded in the copulation-like movements and vocalizations. The performers seem to have "learned"—at least recognized— that the observers understand the new signals. They perform appropriately in the new circumstances. And the partners—the mounted individuals—have become disposed to cooperate with the performers, if only by lifting the tail.

(A word of explanation. Phrases such as *have learned* and *come to understand* are not used carelessly or inadvertently in the preceding passage. Without some degree of mutual comprehension among the participants, the whole interaction would dissolve into real, not only mathematical, chaos. This has been averted, somehow, in most cases.)

The evolutionary history of choking probably has been rather different from that of pseudocopulation. The nesting components, although parts of a display that is unmistakably hostile in large part, do occur during the breeding season. One can imagine, therefore, that the originally nonhostile patterns simply began to be performed over a gradually increasing time span as they began to be integrated with the other parts of the display and, presumably, to acquire some hostile valence of their own. Observers, performers, and partners must still have had to accommodate, to respond correctly to the changes, but they probably could start more slowly and proceed more deliberately than individuals confronted with more abrupt intrusions.

Another process that may have been slow was the development of the penile display of the female spotted hyena. The pseudopenis itself must have taken time to evolve. What the accompanying or subsequent changes in awareness, understanding, or learning must have been are enough to beggar the imagination. Briefly put, the process may be supposed to have involved a switch from male sexual behavior to male hostile behavior, a threat or an assertion, then a transfer of the assertive behavior to the female, followed by a change from threat to appeasement in the male (and other subordinate individuals). Even this may be oversimplified.

East et al. (1993), following Frank et al. (1991), also suggest that high levels of the sexual hormone androstenedione before and after birth may favor both siblicide, with or without maternal approval, and perhaps the general social dominance and virilization of females, with all its behavioral consequences.

We seem to have left structuralism behind and passed on to deconstruction. A serious passage can be quoted: "If the effect is what caused the cause to become a cause, then the effect, not the cause, should be treated as the origin" (Culler 1982, more or less after Nietzsche and Derrida).

Actually, we do not really have to go this far at our level of argument. We might do better to regress a little. W. J. Smith's (1965) distinction between "message" and "meaning" in communication is still useful. The message of a signal is the sum of information encoded within it. The meaning is the information perceived by the observer. The perception is derived from a reading, partial or complete, of the message in its actual context. Context is important. Circumstances are never exactly the same on more than one occasion. Every message that is read more than once is bound to have at least two, several, or many meanings (also see W. J. Smith 1985). When suitably digested and edited, meanings produce results, changes in the attitudes or activities of the observers. In other words, meanings determine the functions of messages, just as the results or functions determine (reveal) the significance of the meanings. This is where the deconstructivist bit begins to loom on the horizon. In any case, it is no wonder that signals often have multiple and diverse effects.

Signals can be broken down into components. It may be convenient to distinguish the message from the format in which it is imbedded. Formats obviously change over time. Presumably, messages are also mutable. They may last while they last. The question is: are the changes in messages and formats closely correlated? As far as I can tell, there are no practical or theoretical reasons why they should always go hand in hand. Yet they might well do so with appreciable frequency. I should like to suggest, in fact, that the messages, not only the meanings, changed during the evolution of some of the most complex displays, the extreme diversions, the strongest controls of aggression. Since the motivation changed, and the effects (the meanings) changed, it is difficult to believe that the crucial intermediate links, the messages, were not also altered or transformed.

Of course, the evidence is circumstantial, and the argument is hypothetical. Earlier suggestions that some present-day patterns are indicators of past conditions were also speculative. Arguments of this type are dangerous. They may allow too many degrees of freedom. They can suggest anything and prove nothing. Yet they are almost unavoidable in discussions of the evolution of behavior. They may focus our attention and give us ideas to test. And some of the hypotheses may even be correct.

COMPETITION AND AGGRESSION: SUMMARY AND CONCLUSIONS

Competition would seem to be universal and, therefore, unavoidable. All resources, of time or space or energy, that are available to one organism are potentially available to be used, perhaps profitably, by other organisms. The supply is limited. In a finite world, all resources will be used up eventually.

Thus, at any given time, organisms must scramble or struggle to get their share of whatever they may need in order to survive and reproduce.

No organism has yet been able to outcompete all the others. It is conceivable that the final result of evolution will be an omnicompetent organism, presumably occupying everything. The survival of the really fittest. The prospect still seems to be remote. For our own planet, the end is more likely to be a miserable arthropod scavenging along a frozen strand as visualized by Mr. H. G. Wells.

Failure in competition must be the usual cause of extinction in the ordinary course of events. Catastrophes may occur; but mass die-offs, as supposedly in the late Permian or around the K/T boundary, are relatively rare (Raup 1988). The equilibria, gradual turnovers and replacements, are longer than the punctuations. Even the most convinced and persuasive advocates of the supposedly distinctive nature of macroevolutionary changes (e.g., Gould 1995) do not "doubt the efficacy of accumulated competition and natural selection through time (for such an explanation is too sensible to be always wrong)."

It may reasonably be supposed, therefore, that all individuals are subject to selection pressures for increased competitive abilities. One way to compete successfully is to discourage, manage, or control competitors. Various behavioral devices have been developed to do this. Social reactions are particularly important. Probably the strongest and most widespread class of social controls is aggression.

Like competition itself, aggression is expensive of time and energy. It has the added disadvantage of being dangerous. Attacking individuals may be defeated, injured, or killed. Naturally, in these circumstances, selection will favor the development of precautionary measures. There will be a tendency for attacks to be made as nearly safe as possible. Aggression, itself a control of competition, has to be controlled in its turn.

PARTIAL RECAPITULATION WITH COMMENTARY

The patterns used for control are numerous and varied.

There are, for instance, many simple postures, intention movements, and other performances that seem to express tendencies to advance and/or to retreat. In effect, these probably are often tendencies to attack and/or to escape. They are sometimes accompanied by, or combined with, indications of other kinds of motivation such as friendliness or sex. Most of these physically simple patterns seem to be of low intensity. Even so, the various tendencies may interfere with one another. For example, impeded attack may be redirected upon scapegoats.

This sort of behavior can be extremely useful in several ways. It may contribute to habituation, the waning of responsiveness in the absence of reinforcement. It must facilitate the establishment of dominance hierarchies, formal expressions of status that often obviate the need for actual attacks or fighting. Various elements also are involved in the initiation and maintenance of territorial arrangements and the development of other systems of mutual avoidance and exclusion.

All overt behavior patterns encode and provide information. Every pattern says something about the motivation of the performing animal, and probably also something about the circumstances of the performance. In this sense, it is potentially a signal. The encoded information can be conveyed or transmitted to perceivers. When and if this occurs, the transmission is an act of communication, and the information in the signal is a message. What a perceiver infers from the message, in its own particular circumstances, can be called the meaning. For a student of behavior, it is sometimes convenient to distinguish the message from the format in which it is embedded.

With some possible exceptions, during transitions or transpositions of signals, messages are usually honest. Only their reading may be difficult, especially in cases of mimicry and crypsis. Misreading may occur. It may even be encouraged. But real cheating (purposeful or not) is rare. Even bluffing or braggadocio is minimal.

Simple and straight patterns such as advance or retreat, although they are informative and can function as signals, are not confined to processes of communication. They do many other things as well (or instead, on occasion). They are conventionally said to be unritualized. They can be compared and partly contrasted with other actions and reactions that are said to be ritualized. These are the patterns that have become specialized in form or frequency expressly to convey information, and perhaps in some cases to do nothing else. Ritualized patterns are usually called displays. They tend to be more complex and elaborate in form than unritualized patterns. Naturally enough, they also are more difficult and more interesting to study.

All acoustic patterns are displays. They may or may not relieve internal stress, provide "emotional relief," but their primary objective function can only be communication. In birds at least, vocalizations can express all kinds of motivation, aggression, and other tendencies, at all levels of intensity, in various combinations and permutations. Notes can be discrete or intergrading. They can be uttered in distinctive, often stereotyped, sequences and arranged in highly structured repertories. They can signal probabilities of attack or escape or other activities with great precision. In many cases, they reveal, even advertise, individual identities and physiological states. All this must be useful in the management of social relations. It may be essential for many birds and some other animals.

Olfactory signals range from unritualized to highly ritualized. Some body odors may be only physiological by-products. Excreta are produced and have to be disposed of "anyhow" in some way or another. They still encode information. Communicative functions become more obvious when urine or faeces are spread or deposited by distinctive methods in special places and/or when special glands are developed to produce distinctive odoriferous secretions. Scent-marking can encode information about identity, mood, probabilities of performance, status, social relations, and associated factors such as the distribution and abundance of food and other resources.

Visual communication also includes a great variety of patterns. Many kinds of information can be encoded in all sorts of postures and movements, again ranging from unritualized to highly ritualized.

The different systems have different properties. Both acoustic and visual patterns can be used as short-range or long-range signals or both. They can be started and turned off abruptly. They can be continued, by elaboration or repetition for more or less lengthy periods without necessarily becoming weaker per se. Only perceivers may become habituated. Olfactory signals also can be short-range or long-range. They differ from acoustic and visual patterns in one important respect. They are partly independent of the constraints of time. They can, in effect, signal into the future. A scent mark can survive long after the marker has moved on. It may fade gradually, but it cannot be turned off abruptly at will.

Tactile signals have been little studied. Perhaps all that should be said here is that, simply because of the close approaches and physical contacts involved, they are both particularly risky and perhaps unusually effective in reducing social tensions.

Statements made in different media, conveyed by different channels, can be used to support or reinforce one another. Acoustic signals can be combined with visual signals and/or olfactory signals and/or tactile signals. For many animals, the number of possible combinations is enormous, probably literally countless, by us, even with the aid of modern technology. The result may not be perfect redundancy. It is doubtful if any two signals are ever identical. But it must be a common occurrence for many slightly different versions of a single message to be broadcast simultaneously or side by side. If crypsis is not involved, every effort is made to ensure that signals are detected and read, rightly or wrongly.

Many nonhuman animals have the physical capabilities to discuss many subjects. They do not seem to do so frequently or extensively on most occasions. But, even when their statements are basically simple, they can be conveyed with many inflections and, presumably, remarkable precision. The animals may not say very much, but they seem to say it very well.

Note that the argument is "one-sided." Precision is presumed because its existence would seem to be the most plausible, perhaps the only, explanation of the form of the performance.

Absolute confirmation is often difficult or impossible in the field. Naturally occurring social behavior patterns may be varied or variable because they are responses to multiple stimuli, not only signals from interlocutors but also input from other features of the surroundings that differ from one event to another. It might be fair to say that the responses that we see in the field are often too mixed, too nearly "opaque," to reveal any single one of their causes completely.

This is an old problem in studies of social behavior, at least of many vertebrates and some cephalopods. What is most interesting should be—or can only be—observed in nature. But the results may be very general or "rough." Conversely, what can be studied most finely in the laboratory is not always very interesting.

These difficulties do not preclude comparisons at appropriate levels of resolution. Consider two systems of signals. The visual displays of birds and mammals are not necessarily more numerous or diverse than the vocalizations of birds. They probably are, however, more heterogeneous in origin. They seem to have been derived from a wide variety of sources. Some of these sources may still be identified by direct comparisons of the shapes and forms of existing patterns.

As a preliminary step, it may be convenient to draw a rough distinction between largely "intrinsic" and largely "extrinsic" sources. Some displays seem to be developments of patterns that were already in place, or at least nearby in time and space, when the process of ritualization began. Other displays would seem to have incorporated elements of more distant or extraneous origins.

In species with parental care, birds and mammals, all or most of the numerous friendly and gregarious patterns, so often used to control or canalize aggression, seem to be derived from familial reactions. They include joining, following, and helping, at the nest and elsewhere. These patterns can be shown, by extension, to all sorts of potential cooperators, competitors, rivals, opponents. The results include the formation of groups of different sizes, sometimes very large (e.g., the immense herds of some ungulates in East Africa), of various levels of structural complexity (there may be rivalries, special preferences, and alliances within groups), and of a wide range of compositions or consistencies, from smooth to lumpy or motley (e.g., the mixed flocks of birds so common in the tropics and temperate woodlands).

Gregariousness can be useful for many reasons. At one level, the members of a group may help one another to find food and/or cope with predators. At a more strictly social level, gregariousness may also help to regulate competition in two other ways, not only indirectly by modifying expressions of hostility, but also directly by facilitating, even ensuring, that associates can monitor one another's activities. No individual is likely to get much of a head start in exploiting a resource when and if its companions are watching with interest at the time.

Among the friendly-gregarious patterns, and the derivatives thereof, that may be performed in potentially hostile situations are a cluster of "allos": allopreening, allogrooming, and alloparental and perhaps even alloinfantile behavior, sometimes supplemented by "altruism" (the etymological resemblance is not entirely coincidental). Some of these patterns are used for appeasement or soliciting. They seem to be more or less slightly or strongly ritualized in different cases. They may also represent different stages of transition from friendly to hostile causation (motivation).

The point may deserve some comment. In most of the cases cited here, it has been assumed that, whenever related or apparently homologous patterns occur in both hostile and nonhostile circumstances, it is the hostile that has been derived from the nonhostile rather than the reverse. This is not inevitable in theory. There is no obvious, compelling reason why signals could not go either way during evolution. Still, it is easier to imagine a pattern of grooming, feeding, drinking, nesting, or some other innocuous activity appearing in a hostile context than an attack pattern appearing in an innocuous context. The former might be distracting or diverting—perhaps the end in view—but the latter might be immediately destructive of other activities in progress.

Apparently exceptional cases, such as the aggressive elements in the courtship of gulls and terns, can be explained (away) fairly easily. The courting birds probably are testing one another, their fitness and courage, in some variation of the handicap principle. It may not be coincidental that the suspected movements from hostile to nonhostile are all categorized as sexual behavior.

Probably always available as sources of new displays or display elements are displacement activities, the anomalous patterns produced apparently out of context, epitomes of the extrinsic or intrusive. As it happens, all the supposed displacement patterns that have been described in the literature are visual; most of them are movements. There is no obvious reason why acoustic or olfactory or other patterns could not also be displaced. Perhaps it is only that they have been overlooked by human observers.

Whatever the sources, all displays must change and develop over time. If there is an initial stage of surprise, startle, or distraction, it probably does not usually last for a very long time. Surprise is an effect that is difficult to maintain. The situation needs to be "regularized." In the case of patterns occurring in hostile contexts, there will be selection pressure for them to become clearer in several respects. The motivation of a performer, whatever it may have been before displacement, should become consistently hostile. The performer should be able to call upon the pattern "at will." It should behave as if it knew what it was doing. Perceivers should come to behave as if they understood the hostile valence of the performance. And the performer should come to expect the reactions of the perceivers and react appropriately.

At this point, the pattern will have become an ordinary display *comme les autres.*

It will then begin to encounter the usual difficulties and vicissitudes of most ordinary patterns. It probably will become attenuated sooner or later, losing its impact by continued repetition. Perceivers will become habituated. As far as I know, all stimuli, in all media, are vulnerable to this process, although the details may differ widely in different cases.

However variable, habituation is essentially a short-term phenomenon from a comparative point of view. There also are long-term changes. Displays tend to become old or old-fashioned over periods of hundreds or thousands or even more years. When this happens, evolution may favor exaggeration or rearrangement of the stimuli presented (e.g., the peacock's tail). Some form of "jazzing up." Madder music and stronger wine. Sooner or later, however, an elderly display probably will be reinforced by new elements or replaced by something different. Quite possibly, the new elements and replacements will have originated in the same way, if not from the same sources, as the now senescent pattern in decay. Flux can be progressive or repetitive or both.

There is another factor involved. It can be a simplifying or a complicating influence in the evolution of signal systems. It is, in a way, a matter of "evolutionary economy." A pattern that has been evolved to subserve one function may come to acquire other functions, even very different ones. In many cases, it seems to have been "easier," more often favored by natural selection, to join new functions to old structures than to develop new structures for each and every function. The various functions may or may not coexist indefinitely. The general phenomenon has long been well known. Most examples that come to mind immediately are anatomical or paleontological: the use of a hard shell, originally designed for protection, as a floating device (the earliest cephalopods); the use of paddling fins for locomotion on land (some fishes); the use of jawbones to enhance sound reception (mammal-like reptiles). Behavior can provide many comparable examples: the remarkable histories of grooming, preening, nesting, begging, and sex in hostile contexts.

THE VARIETY OF RELATIONS AMONG AND BETWEEN TACTICS AND STRATEGIES

Most of the behavior patterns in the preceding recapitulation are, in some sense, tactical. They are the behavioral means by which, if all goes well, competition and hostility are controlled, desired ends or goals are achieved. They are not usually the ultimate ends or goals themselves. Ultimate objectives are strategical.

Strategies can be carried out only by tactics. Nevertheless, it is remarkable how flexibly, even loosely, the two levels of adaptation, the two kinds of behavior, are coupled. Or, rather, partly decoupled.

Different aspects of the subject are differentially accessible to analysis. Behavioral tactics are, or should be, relatively easy to study for several practical reasons. They are often available for actual examination. Most of them can be observed in fairly short periods of time. Analyses of long-term strategies, by contrast, are both more difficult and too easy. One cannot watch the whole of a strategy unfold before one's very eyes. One can only see bits (bites) and pieces. There has to be recourse to extrapolation. Therein lies a danger.

Consider a purely hypothetical example. Let us suppose that we are told that species A lives in habitat X and is monogamous, while related species B lives in habitat Y and is polygynous. Given these data, any behaviorist worth his or her salt can develop a theory or scenario to explain why monogamy has been selected for in the first case and polygyny in the second case. But then, suddenly, it is discovered that some of the supposed data are wrong, have somehow been transposed by mistake. It is A that is polygynous and B that is monogamous. Some ingenious mind will quickly explain why polygyny was selected for in the first case and monogamy in the second case. *Caveat lector.* Heads I win; tails you lose.

In these arguments, as in some others, there may be too much freedom of choice in suggesting answers. This must be guarded against. Field biologists may have to proceed as if they believed that "whatever is, is right"; but they should cultivate a very strong dose of skepticism at the same time.

Perhaps the most nearly reliable method or technique for historical reconstruction of social behavior is comparative analysis of the behavior of living forms. This is the good old method of the classical ethologists, Heinroth (1911), Whitman (1919), as well as Lorenz (1931, 1935, 1939, 1941). Most reliable, but not subject to absolute proof.

Doubt should not preclude further discussion. With all due skepticism, it would still seem to be safe and correct to make two generalizations, which should apply to behavior as well as to other aspects of life: (1) *The same or very similar tactics subserve different strategies* in different cases. (2) *Different strategies are used to solve the same or very similar problems* in different cases. A third possibility, that different tactics can subserve the same or very similar strategies, should perhaps be considered.

The first point was mentioned earlier. It certainly is valid at some level. Advance, retreat, threat, appeasement, transferred sexual or parental patterns, and so on can all be used to form, regulate, and maintain all sorts of social relations, large groups or small groups, egalitarian bands and rigidly hierarchical societies, exclusion or avoidance, eusociality, promiscuity, polygyny, polyandry, what have you.

It is not the nature or identity of the basic elements of social behavior that determine the outcome. Rather, it is the ways in which the various elements are used. Where, when, why, by whom, to whom, with what frequencies, in which circumstances?

Our knowledge of strategies is less complete, at least less precise, than our knowledge of tactics. In particular, there have been few accounts of aggressive or hostile strategies as such. Still, something can be learned, usually by inference, from discussion of feeding and sexual strategies. These include hostile patterns as well as other actions and reactions. Thus, for instance, it must be supposed that the hostile behavior of solitary bee-eaters differs, at least on average, from the corresponding behavior of the more gregarious, communally breeding types. If nothing else, there must be differences in average frequencies of performances of such patterns as overt attack, escape, appeasement, redirection, and diversion. Fry (1984), following Lack (1968), suggests that the differences in social organization of different species are adaptations to different diets, different spatial and temporal distributions of different foods (insect prey). In a rather similar argument, Halliday (1980) suggests that the polyandry of jacanas is useful because it enables females to feed more frequently, more nearly continuously, in order to produce more eggs, perhaps necessary to compensate for frequent destruction of nests.

The published literature would seem to imply that hostile strategies are somehow dependent upon other social and ecological adaptations. This may well be true, up to a point, as a general rule, with possible exceptions. Strategies are less easily arranged and rearranged than tactics in some, but not necessarily all, circumstances.

Perhaps the most conspicuous example of decoupling cited in this book is the contrast between tree sloths and leaf-eating monkeys. The two types of folivores occupy similar, probably widely overlapping, ecological niches in the same areas. Both are very successful. Their populations are large and dense. Yet their general social strategies could hardly be more different. Their hostile repertories are particularly distinct. The tree sloths show little more than avoidance and exclusion. In the end, however, the different strategies of the two types seem to be approximately equally effective.

It is these sorts of apparent inconsistency or incoherence that have suggested to some students that evolution is "opportunistic." Again, this may well be true, but only up to a point. Serendipitous accidents may occur. Nevertheless, it should be assumed, simply for methodological reasons, economy of hypotheses, that accidents do not usually remain accidental for very long during evolution. Mutant characters only become "right" when they are positively selected for.

A MAIN LINE? LEAD, KINDLY LIGHT

However opportunistic evolution may be, it is not a random walk. It moves in directions. Some directions may be traced through the sometimes tangled thickets of social behavior.

The traces are not always obvious. The various signals and controls used in hostile interactions are exceedingly numerous. Relations among them often are intricate. Lines of functional and/or phylogenetic descent drawn in diagrammatic form would show many crotchets, zigs and zags, reversals and crossovers (all the peculiarities noted above, and more). It seems to be possible, nevertheless, to trace something like a rough and obviously incomplete "main sequence" in fairly objective terms as follows:

a. Simple unritualized advance and retreat movements and intention movements, alone or in alternation or in combination. During hostile encounters, advance and retreat are likely to be expressions of attack and escape. When used appropriately in the right circumstances, these patterns are quite capable of maintaining "public order" with a minimum of overt violence.

b. Ritualized attack and escape movements. These are very much like the patterns of the preceding group, but definitely stereotyped and sometimes exaggerated in form or frequency. (Actually, I was struck, on rereading the behavioral literature, by how frequently a regular change of orientation is one of the first signs of standardization.) These lightly ritualized patterns also help to regulate social relations with little or nothing in the way of actual fighting. Only they are often accompanied by a certain air of tension. They can become ceremonious, stiff and self-conscious looking.

Vocalizations may occur with patterns of both the a and b groups, perhaps more often with b than with a. Certainly more frequently by many birds than by most mammals. Such as they are, the vocalizations are not usually very long, loud, or complex, at least in purely hostile situations at this level. (The more ambivalent songs and other signals to attract mates as well as to repel rivals are another matter.)

c. Diverse displays, diversions, without physical contact. Many hostile displays, especially performances of some appreciable morphological elaboration, include components of nonhostile origin. These extrinsic or extraneous elements come in many shapes and kinds. They are derived from many sources. For our purposes here, it may be convenient to distinguish, as a recognizable group or class, those displays whose incorporated elements are derived from "innocuous" sources, for example, nesting, feeding, drinking, or preening (self-preening). These sources can be said to be innocuous because they do not entail physical contact between performers. This is still true of their hostile derivatives.

d. Diverted displays with physical contact. These are similar to the patterns of the preceding group except for the fact that their incorporated components are derived from allopreening, allogrooming, copulation, parental-infantile interactions, and perhaps other elements that definitely do entail physical contacts. It seems likely that such contacts are always potentially dangerous. If and when fighting should develop, accidentally or in the normal course of events, proximity must increase exposure and vulnerability.

An interesting point is that some of these patterns are done *to and with* allies in the face of enemies (the pseudocopulations of some blue-bellied rollers), or to and with antagonists that might become allies in the future (the pseudocopulations of pied kingfishers, the dominance mountings of many Old World monkeys and apes, the allogrooming of some of the same and related species).

Individuals performing pseudosexual or allogrooming patterns may be taking risks. They may be conspicuous while doing so. They are not, however, necessarily demonstrating their willingness and ability to overcome handicaps. Or, at least, this may not be the only thing that they are doing. By breaching rather formidable social barriers, they may (also) be encouraging the development of mutual trust and confidence: from the hostile back to the nonhostile.

These four groups are recognized primarily on the basis of movements and postures. As signals, they are primarily visual. There is not much to be said, in this limited context, about other sensory modalities. Acoustic signals occur with c and d as well as with a and b. Olfactory signals of one sort or another probably occur in the course of many different kinds of conflicts. Tactile signals, almost by definition, are characteristic of d.

There may be another category: explosion. Nature red in tooth and claw. There are encounters in which fighting gets out of control, to no useful purpose. This may be accompanied by displays in the intervals, usually in broken, disjointed, hectic, or frantic forms.

The series a to d is a sequence. It may be a progression. But from what to what or where to where? I would like to suggest that it indicates or expresses increasing *intensity* of conflict. Not physical combat, *but rather conflict between or among different kinds of motivation, tendencies to do different things.* The crux seems to be the *relative* or comparative, not (or not only) the actual, strengths of the tendencies involved. Thus, for instance, activation of an attack "drive" at any level, high or low, simply leads to the performance of attack movements, strong or weak, in the absence of any contrary or counteracting tendency of some appreciable force. In the case of attack, a tendency to escape must be the most frequent counteracting factor. There must be other or additional factors on particular occasions. It is when incompatible tendencies are similar in strength, at least above a minimal level, that complex performances such as displays are most likely to appear. Complexity may increase progressively as motivation becomes progressively stronger, but only if and when the relevant tendencies remain "balanced," roughly equivalent to one another.

THE CONTEST AND THE MORAL

At this stage of the argument, it may be appropriate to resort to metaphor and simile. Even ethologists may find references to games useful for abbreviation or emphasis.

The history of interactions between aggression and devices to control aggression must be very long. Given the material, one is reminded of an arms race, doubtless escalating over time. But the race is not exactly symmetrical. The runners, the two sides, are different in quality. Perhaps the interactions between aggressive patterns and their controls (escape, appeasement, ingratiation, diversion) should be compared with other phenomena such as the race between mammalian ungulates and carnivores to develop better brains (references in Cifelli 1985), or the escalation of predatory and protective devices among marine animals (described in detail by Vermeigh 1987).

Better yet may be a comparison with games of cards. Not poker. Bluffing is not greatly encouraged, and the pot is seldom bought. The more complicated and modulated game of bridge may be a more useful analogy (as in some other cases: see Moynihan 1982a).

Consider what happens during hostile interactions. The participants (no specified number) are dealt "hands." They are given opportunities to "play," to behave in certain ways. The hands consist of "cards" of various "suits." During hostile encounters, the suits are strategies such as aggression, avoidance, or appeasement. Although they are diverse, they are relatively few. The cards can be considered tactics. At least at the beginning of an interaction, they usually are more numerous than the suits.

Someone opens the game by playing a card. Then the other participants play their own cards. Animals in the real world do not necessarily play in very ordered sequences. Nor do they necessarily have to follow suit. Eventually, however, one card proves to be "stronger" than the others. The owner may be said to have won a "trick." Play may stop after one or more tricks have been taken. The break may be permanent or temporary. Contests may be renewed. Repeated encounters or "rubbers" can be cumulative.

The results must depend upon many factors. The same participants may be dealt different hands at different times. Different participants may be

Fig. 14. The tale ends. Tail-twining by titi monkeys *(Callicebus moloch)*, probably for reassurance and as indication of a happy relationship. From Moynihan 1966, 83, fig. 1.

dealt the same or similar hands. Some participants are luckier or more skillful than others.

There may be equivalents of preemptive bidding or false-carding or even revoking. But they cannot occur too frequently. Otherwise the system would fall apart—which it obviously has not done.

Argument and analogy are concordant.

There must have been many losers of contests in the past. There also have been many winners. These are the types that have survived to the present. It must be supposed that the winners have won, so far, because they evolved both appropriate strategies and effective tactics, or arrangements of tactics, to pursue these strategies successfully. Their bids have been good, and their play has been at least adequate.

The importance of details, the components of behavior, cannot be overemphasized. It is a truism of military history that wars, like other contests, are decided by the accumulation of skirmishes (citations from Clausewitz and others in Turner 1985).

The point may be generalized to many endeavors. Regardless of epigrams, it is not only the first step that counts. All the steps that follow after are also determining (fig. 14).

Appendix 1

SCHEDULES OF OBSERVATIONS OF CORACIIFORMS

I actually looked at, or for, coraciiforms in the field on the dates cited below. In the regions visited most frequently, West Africa and southern India, I was able to cover a variety of habitats at different seasons. Supplementary observations in the field were made in central Madagascar in the forest reserve of Perinet, at Bharatpur in Rajasthan, and in Queensland on Lizard Island and near Townsville. Captive individuals were observed in the Shanghai Zoo and the United States National Zoo's Conservation and Research Center at Front Royal in northern Virginia.

West Africa

Sénégal, mostly in the Casamance, along the Petite Côte, and on Cap Vert: Aug. 14-Dec. 9, 1976; Jan. 22-Mar. 12, 1977; June 15-21, 1981; Sept. 22-Dec. 2, 1985; Sept. 13-20, 1989; Oct. 6-18, 1991.

Gabon, at Makakou: Dec. 17-28, 1976; the Ivory Coast near Abidjan: Jan. 1-3, 1977; Liberia, on Mount Nimba: Jan. 5-19, 1977.

Southern and Western India

Mostly in Tamil Nadu near Mahabalipuram, with a few brief excursions to Kerala and Goa: Dec. 18, 1978-Jan. 15, 1979; Mar. 7-9, 1980; Oct. 17-20, 1981; Apr. 4-14, 1984; Jan. 5-10, 1987; Sept. 22-23, 1990.

Orissa

Mostly near Bubaneschwar: Jan. 13-18, 1987.

Nepal

Chitwan Park: Apr. 18-29, 1984.

Southern Asia

Peninsular Malaysia: Jan. 19-30, 1979; Singapore: Mar. 5-8, 1979; Sumatra: Feb. 2-14, 1979; Java: Feb. 18-Mar. 1, 1979.

Burma: Oct. 24-26, 1981; Yunnan: Oct. 30-Nov. 3,1981.

South Pacific

Guam: Mar. 23, 1979; Palau: Mar. 15-20, 1979; July 19, 1981.

New Guinea

North coast near Jais Aben: Oct. 17-27, 1986; near Port Moresby in the south: Oct. 29-30, 1986.

Appendix 2

CORRELATES AND CONSEQUENCES OF NESTING HABITS

More or less conventional opinion, at present, suggests that living birds include something like 9,000–9,500 species. Of these, something like 58–59 percent are supposed to belong to the order Passeriformes—the passeriforms, or (as they are sometimes loosely called) the perching birds. Several other systematic groups, perhaps other orders, are sometimes associated in systematic lists with the passerines and with each other under the general term of "near passerines." They are supposed to number only a few hundred species.

The figures and estimates cited above are derived from Sibley and Ahlquist (1990), who list their sources in detail. There is an even newer school of classifiers, taxonomists, who would recognize many more species (references in Martin 1996; and Zink 1996). They do so mostly by raising bird populations from lower ranks such as subspecies to full species rank. Thus, the difference between the two arrangements is partly nominal. In any case, the relative preponderance of passerines in number of species is as great in the latest classification as in any of the earlier ones.

The passerines include such familiar types as larks, warblers, thrushes, flycatchers, titmice, jays and crows, finches and sparrows, as well as many more exotic forms of the various regions of the tropics. As a rather vaguely defined concept, the near passerines may be supposed to include the

Coraciiformes already described; plus the Piciformes, barbets, honey-guides, toucans, and woodpeckers, possibly also the jacamars and puffbirds; the Coliiformes, the colies or mousebirds; the Cuculiformes, cuckoos and perhaps turacos; the hummingbirds of the family Trochilidae; and perhaps swifts (Apodidae).

Many of the birds of these groups are similar to one another in aspects of "habitus" or way of life. They tend to be small or smallish (the large hornbills are the most conspicuous exceptions). They are basically insectivorous or omnivorous with, in some cases, excursions into nectarivory, frugivory, or even predation upon small vertebrates. They perch on all sorts of supports, stones, stalks, twigs, branches, etc. Most species can also use their legs and feet for standing, hopping, striding or running on the ground.

Given the similarities between passerines and near passerines, it might have been expected that they would have had similar evolutionary "potentials." Obviously, they have not fulfilled this expectation. The species numbers reveal that the groups have been differentially successful in some respect(s). What respect(s)? And why?

There are some anatomical differences between groups (in fact, the criteria by which the various groups have been recognized by systematists.) Yet these differences would appear to be relatively minor from a large-scale evolutionary point of view. Or, at least, none of them is obviously *preclusive.* None of the distinctive features of most near-passerine groups seems to be so closely adapted to one particularly narrow niche or way of life as to prevent the evolution of new characters to permit advances in new directions and the exploitation of new opportunities.

As far as I know, the physiological and behavioral characters of all these birds also are "open" to change.

Approximately four-fifths of the living species of passerines have a specialized or derived type of vocal apparatus, a diacromyiodian syrinx (see Ames 1971). They are called "songbirds" or "oscines." The particular advantages of the specialization of the vocal structures are not clear from the published literature on the subject. Songbirds are not the only birds that sing. It is true that they can utter sounds in intricate combinations. But so can many other birds, some woodhoopoes and hornbills, for instance.

Passerines do not necessarily utter or rely upon their vocalizations more frequently than many nonpasserines. Some barbets and cuckoos, the "brain fever birds" of the Old World tropics, are famous for the long repetitions of their calls or songs. Monotonous as such performances are to human ears, they must be valuable to the performers.

Many species of oscines learn parts or aspects of their songs. This learning may depend upon special structures and processes in the CNS (references in H. Williams and Nottebohm 1985). Some hummingbirds of the genus *Phaetornis* also learn their songs (E. S. Morton, pers. comm.). There is no suggestion that songbirds as a group (apart from a few special cases such as the corvids) are generally more intelligent than their relatives.

Perhaps the distinctive features of the vocal systems of oscines are as much a product as a cause of their evolutionary success.

There is another factor to be considered. By and large, passerines and near passerines differ from one another in their nesting habits. The nonpasserines, with exceptions, nest in holes. The passerines, again with exceptions, construct nests outside of holes. Such nests can be said to be built in the "open," although they are often masked or hidden by vegetation. However concealed, they probably are, on the average, more vulnerable to predators than are clutches of eggs in holes. They can be damaged by wind and rain. They must have compensatory advantages. They can be built of a great variety of materials at a great variety of sites, anywhere that supports can be found. The subject has been reviewed by Oniki (1985).

Since hole-nesting is characteristic of most of the different "orders," it probably is primitive in the near-passerine-passerine series as a whole.

In their nesting habits, most passerines would seem to have followed a high-risk strategy. They have, thereby, attained a degree of freedom and flexibility. It probably is simply because they can nest almost anywhere that they have been able to occupy and utilize more of their habitats, in a greater range of environments, than their hole-nesting relatives. This can hardly have failed to facilitate, perhaps stimulate, the multiplication of species, at least indirectly in the medium to long term.

The multiplication itself must have had consequences, certainly for social relations among species. There must have been selection pressures for refining and differentiating signal systems, including acoustic patterns. Acquisition of new patterns by learning rather than mutation would speed the development of new methods to cope with more players, respondents and correspondents.

What is being suggested here is that the original innovation, the "invention" or "trick," primarily responsible for the evolutionary success of the Passeriformes was neither perching nor singing, but rather the habit of building nests in the open, outside of cavities.

Some of the exceptions noted above are suggestive. Thus, for instance, the hummingbirds are the most speciose of the near passerines. They include more than four hundred species according to Greenewalt (1960). Besides learning their songs (in at least some cases), they also construct their nests. They would seem to be following the same path as the passerines.

The coraciiforms have not been so innovative. Tied down to their hole-nesting habits, they have played safe—with very considerable success, but only within certain limits.

GLOSSARY

PREPARED BY OLGA F. LINARES

Some general definitions are here supplemented by usages specific to the author of this book.

ADAPTIVE: traits that are useful and/or advantageous in particular natural and social environments. They were selected for because they resulted in relatively better survival and/or reproductive success.

AGGRESSION: intraspecific and interspecific behavior patterns that serve to intimidate or damage another organism, ranging from violent blows or strikes to simple intention movements. The term also includes "mixed," ambivalent performances in which elements of attack can be detected or inferred. Aggression seems to be primarily an adaptation to cope with competition.

AGONISTIC: aggressive patterns, with or without escape elements. Here, I will generally use *hostile* instead of *agonistic,* but the two terms are virtually synonymous.

ALLOINFANTILE BEHAVIOR: reacting to other individuals as if they were one's own parents when they are not.

ALLOPARENTAL BEHAVIOR: taking care of other individuals, usually young, as if they were one's own offspring when they are not.

ALLOPREENING OR ALLOGROOMING: maintenance activities that are transferred, as when one individual preens or grooms another instead of itself. The first is characteristic of birds; the second of mammals.

ALLOSEX OR PSEUDOSEX: behavior of sexual copulatory origin performed (usually as signals) in relations and situations that are not directly or immediately sexual at the time of performance. Its possible functions include dominance assertion, tension regulation (lowering), reconciliation, and alliance formation.

ALTRUISM: the term usually implies some real sacrifice of inclusive fitness by the giver of the benefit on behalf of a conspecific.

APPEASEMENT: any behavior pattern that reduces an opponent's tendency to attack without, at the same time, greatly increasing its tendency to escape. Appeasement can occur in a wide range of social circumstances. It is common in intraspecific encounters.

AVOIDANCE: successive use of resources among actual or potential competitors who occupy the same or broadly overlapping territories more or less continuously. Staying away may be long and/or short term, and occur again and again. As in exclusion, factors of aggression, restraint, and a willingness to retreat seem to be involved. Avoidance is especially common during interspecific competition.

BADGES: see *handicap principle.*

BEGGING: behavior patterns used by young animals, for instance birds, to induce, even force, their parents to feed them.

COMMUNICATION: the conveyance of information from one individual to another mediated by acoustic, olfactory, tactile, visual, chemical, and other signals, emitted in social interactions, mostly with conspecifics.

COMPETITION: occurs among animals whenever one individual occupies or preoccupies a resource that would otherwise be available to, and possibly or probably appropriated by, another individual of the same or another species. Two processes have been called "interference competition" and "exploitation competition" (Maurer 1984; Merila and Wiggins 1995; Minot 1981; Minot and Perrins 1986). The first is more or less direct. It may include face-to-face encounters between competitors. The second is more or less indirect. The resources in play or at risk are disputed by long-term or long-distance maneuvers. The two processes are different, but not always distinct.

CONFLICT: as used here it usually or often implies the performance or imminent probability of actual fighting. The term *contest* may be preferable as a broader or more general label.

COTERIES: smaller subgroups of larger colonies of animals.

DIMORPHIC: when males and females are distinguished from one another by possession of different characters that are both permanent and noticeable. The particular characters distinguishing the sexes are different in different groups and species.

DISPLACEMENT BEHAVIOR: behaviors that are "out of context"; for example, in avian behavior, preening of the back or scapular feathers during courtship, for instance, or pecking at the ground in displacement feeding. Single displacement patterns and many bouts of displacement tend to be brief, but they can also be quite elaborate. The frequency of displacements is different in different animals.

DIVERSIONS: the appearance of an old pattern in a new social context. As new displays in their new circumstances, diversions could have appeared abruptly or gradually, by a slow slippage or reorientation in time (rather than space). Such patterns may work at first by distracting antagonists and/or positively stimulating nonhostile feelings or responses.

EUSOCIALITY: in insects, only one female, the alpha female or queen, breeds successfully in any given group or colony with a large number of nonreproductives taking care of the offspring. In mammals eusociality occurs in dwarf mongooses and mole rats.

EXCLUSION: potential competitors occupying different and nonoverlapping, if often adjacent, areas for long periods of time. The separate areas may be defended as territories. The mechanics of exclusion appear to be almost entirely behavioral. Exclusion is common during intraspecific competition.

GAME THEORY: a concept in applied mathematics "dealing with situations in which the 'winning' and 'losing' vary with the strategies chosen by the players" (Immelman and Beer 1989, 116). Game theory itself is a serious discipline. It is explained in rigorous mathematical terms by Straffin (1993). As invoked by sociobiologists, however, it is not always so rigorous.

GREETINGS: commonly applied to ritualized ceremonies or forms of encounter. Almost by definition, greetings occur when individuals join or rejoin one another.

GREGARIOUSNESS: living in closely integrated groups whose cohesion depends upon some developed form of visual, acoustic, tactile, and/or olfactory communication. Associated behaviors include approaching, joining, following, and other friendly patterns considered to be distance decreasing.

HABITUATION: the attenuation or decreasing responsiveness to repeated or continuous stimulation. If and when it is achieved, habituation may lead to a decrease, even disappearance, of aggression, resulting in peaceful coexistence or even cooperation.

HANDICAP PRINCIPLE: an animal may carry or show off some character that might have been supposed to be burdensome, a handicap, simply to demonstrate or advertise that it is strong and fit enough to carry the burden without suffering. The burdens are often called *badges.*

INCLUSIVE FITNESS: the reproductive success of an individual, or the survival of its genes in the offspring of its relatives.

INFANTICIDE: the killing, active or passive, of young, including the active killing of the offspring of potential mates; the active killing of miscellaneous young, including some who are not the offspring of potential mates; and the killing—or letting die—of one's own offspring.

INTENTION MOVEMENTS: foreshortened performances that express tendencies to act, such as advancing and/or retreating.

KIN: individuals that are closely related to one another by descent.

KIN SELECTION: "Within a population, increase of genes that cause individuals to promote the survival and reproductive success of relatives" (Immelman and Beer 1989, 165). Kin selection was and is significant in the evolution of many kinds of social organization, including those that are not openly gregarious or cooperative. It does not, however, explain everything.

MESSAGE and MEANING: the message of a signal is the sum of information encoded within it. The meaning is the information perceived by the observer from a reading, partial or complete, of the message in its actual context. Meanings determine the functions of messages; functions determine (reveal) the significance of the meanings.

MIGRATION: the regular moving between places relatively far apart. The scale is different from that of local avoidance. Migration often results in the removal of potential competitors for appreciable periods of time. Competition and aggression may contribute to the mix of selection pressures in favor of migration.

PLAY: a pleasurable imitation of serious activities. In older animals its function may be to placate potential opponents.

PRISONER'S DILEMMA: "A situation in which it pays each of several . . . agents individually to behave in a particular way, even though it would pay them as a group to behave in some other way" (Bannock et al. 1991, 30–34). Essentially, a prisoner must decide to cooperate with or defect from (cheat) fellow prisoners. The decision is strategic, dependent upon circumstances. The concept is central to game theory.

PROSOCIAL and ANTISOCIAL BEHAVIOR: friendly and unfriendly behaviors that are often associated, even combined or mixed. It could be argued that all interactions among individuals (apart from predator-prey encounters) are social in some broad sense.

REDIRECTION: used in different senses by students of behavior. Here it is used in what seems to have become the conventional ethological sense. It is said to occur when aggression released by one stimulus is vented upon some other organism or object apart from the stimulus.

REGULATION: an individual maximizing good effects and minimizing bad effects to enhance and refine its own competitive abilities while at the same time weakening or controlling the competitive efforts of others. Regulation can only be managed by or through behavior, action and reaction.

RETREAT: to try to play safe by removing oneself from the scene to a nonthreatening distance. A cautious or reluctant contestant can be more or less pacific, even ingratiating, or openly "cowardly." Bilateral and multilateral retreats can be repeated again and again after interruptions.

REVERSE MOUNTINGS: copulatory-type behavior in which the original mounter slides off and is mounted in turn by the individual that had been mounted earlier. "First A on B; then B on A . . . both the individuals involved played both male and female roles in rapid succession." Effectively a kind of defensive threat.

RITUALIZED PERFORMANCES: patterns that *have* become specialized in form and/or frequency *expressly* to convey information. Ritualized patterns usually are called *displays.* Patterns whose communication effects are in some sense "incidental" to their other functions are often said to be unritualized.

SELECTION PRESSURES: pressure exerted by features of the natural environment, including conspecifics and other animals, or an element in mate-choice, that results in changes in gene frequencies within a population.

SIGNS and SIGNALS: behavior patterns that as instruments of conveyance encode information. Almost anything and everything is potentially capable of conveying information and therefore of acting as a signal, even when communication is not the primary function.

SOCIAL MIMICRY: resemblances (such as of color) that have become exaggerated as a special adaptation to facilitate associations of different species. Social mimicry depends upon several factors, including spatial parameters and individual distances, factors that affect the frequencies with which individuals are brought face to face with one another.

SOLICITING: behaviors, such as begging or quivering, that may release or stimulate copulation. They often seem to reduce probabilities of fighting between potential or actual mates.

STRATEGIES: ultimate objectives of behaviors, carried out only by tactics. The same or very similar tactics subserve different strategies in different cases. Different strategies are used to solve the same or very similar problems in different cases.

TACTICS: the behavioral means by which, if all goes well, competition and hostility are controlled, desired ends or goals are achieved. They are not usually the ultimate ends or goals themselves.

TENDENCY: probability of performance (the equivalent of *drive* in some of the older literature). Stimuli that activate a tendency to attack usually release a tendency to escape as well.

VOCALIZATIONS: acoustic performances of all animals (apart from "incidental" noises such as rustling and other locomotory sounds) that can be assumed to be displays in this sense.

References

Bannock, Graham, R. E. Baxter, and E. Davis. 1992. *The Penguin Dictionary of Economics.* 5th ed. London: Penguin Books.

Immelman, Klaus, and C. Beer. 1989. *A Dictionary of Ethology.* Cambridge: Harvard University Press.

MacFarland, D. *Oxford Companion to Animal Behavior.* Oxford: Oxford University Press.

BIBLIOGRAPHY

Agrell, J. 1995. A Shift in Female Social Organization Independent of Relatedness: An Experimental Study on the Field Vole *(Microtus agrestis). Behavioral Ecology* 6:182–91.

Alexander, R. D. 1974. The Evolution of Social Behavior. *Annual Review of Ecology and Systematics* 5:325–83.

Ali, S., and S. D. Ripley. 1970. *Handbook of the Birds of India and Pakistan.* Vol. 4. Bombay: Oxford University Press.

Ames, P. L. 1971. The Morphology of the Syrinx in Passerine Birds. *Bulletin/Peabody Museum of Natural History, Yale University* 37:1–194.

Anderson, D. J. 1995. The Role of Parrots in Siblicidal Brood Reduction of Two Booby Species. *Auk* 112:860–69.

Andersson, M. 1976. Social Behaviour and Communication in the Great Skua. *Behaviour* 58:40–77.

Angst, W. 1974. *Das Ausdrucksverhalten des Javaneraffen Macaca fascicularis Raffles.* Advances in Ethology 15. Berlin and Hamburg: Paul Parey.

Appert, O. 1968. Zur Brutbiologie der Erdracke *Uratelornis chimaera* Rothschild. *Journal für Ornithologie* 109 (3): 264–75.

Archer, J. 1988. *The Behavioural Biology of Aggression.* Cambridge: Cambridge University Press.

Ardrey, R. 1966. *The Territorial Imperative.* New York: Dell Publishing Co.

Baker, E. C. S. 1930. *Gamebirds of India, Burma and Ceylon.* Vol. 3. London: John Bale and Son.

Barlow, G. W. 1985. Reproduction in Reef Fishes. *Transactions of the American Fisheries Society* 114:314–15.

Barlow, G. W., and T. E. Rowell. 1984. The Contribution of Game Theory to Animal Behavior. *The Behavioral and Brain Sciences* 7:101–3.

Barlow, G. W., and J. Silverberg (eds.). 1980. *Sociobiology: Beyond Nature/Nurture?* AAAS Selected Symposium 35. Boulder: Westview Press.

Bateson, P. 1995. Reply from P. Bateson. *Trends in Ecology and Evolution* 10:83.

Bayer, R. D. 1982. How Important Are Bird Colonies as Information Centers? *Auk* 99:31–40.

Beachly, W. M., D. W. Stephens, and K. B. Toyer. 1995. On the Economics of Sit-and-Wait Foraging: Site Selection and Assessment. *Behavioral Ecology* 6:258–68.

Beebe, W. 1918–26. *A Monograph of the Pheasants.* London: Witherby.

Beehler, B. M., T. K. Pratt, and D. A. Zimmerman. 1986. *Birds of New Guinea.* Princeton: Princeton University Press.

Behaviour. 1994. Vol. 130, pts. 3–4.

Bekoff, M. 1995. Play Signals as Punctuation: The Structure of Social Play in Canids. *Behaviour* 132:419–29.

Bell, H. L. 1981. Information on New Guinean Kingfishers. *Ibis* 121:51–61.

Bennett, A. T. D., I. C. Cuthill, and K. J. Norris. 1994. Sexual Selection and the Mismeasure of Color. *American Naturalist* 144:848–60.

Bennun, L. A., and A. F. Read. 1988. Joint Nesting in the Acorn Woodpecker. *Trends in Ecology & Evolution* 3:319.

Berger, J. 1986. *Wild Horses of the Great Basin.* Chicago and London: University of Chicago Press.

Biadi, F., and P. Mayot. 1990. *Les faisans.* Paris: Hatier Editions.

Biggins, J. G. 1984. Communication in Possums: A Review. In *Possums and Gliders,* ed. A. P. Smith and J. D. Hume, 35–57. Australian Mammal Society. Chipping Norton, NSW, Australia: Surrey Blatts and Sons PTY.

Björklund, M. 1984. The Adaptive Significance of Sexual Indistinguishability in Birds: A Critique of a Recent Hypothesis. *Oikos* 43:414–17.

Blaffer Hrdy, S. 1977. *The Langurs of Abu.* Cambridge: Harvard University Press.

Boag, D. 1982. *The Kingfisher.* Poole: Blandford Press.

te Boekhurst, I. J. A., and P. Hogeweg. 1994. Self-Structuring in Artificial "Chimps" Offers New Hypotheses for Male Grouping in Chimpanzees. *Behaviour* 130–52.

Boinski, S. 1994. Affiliation Patterns among Male Costa Rican Squirrel Monkeys. *Behaviour* 130:191–209.

Boinski, S., and C. L. Mitchell. 1994. Male Residence and Association Patterns in Costa Rican Squirrel Monkeys. *American Journal of Primatology* 34 (2): 157–69.

Boissy, A. 1995. Fear and Fearfulness in Animals. *Quarterly Review of Biology* 70 (2): 165–91.

Bovet, G. 1931. *Faune Tropicale 17: Oiseaux de l'Afrique tropicale.* Vol. 2. Paris: Office de la Recherche Scientifique et Technique Outre-Mer.

Bradbury, J. W. 1981. The Evolution of Leks. In *Natural Selection and Social Behavior,* ed. R. D. Alexander and D. W. Tinkle, 138–69. New York: Chiron Press.

Brosset, A. 1983. Parades et chants collectifs chez les couroucous du genre *Apaloderma. Alauda* 51:1–10.

Brown, E. D., S. M. Farabough, and C. J. Veltman. 1988. Song Sharing in a Group-Living Songbird, the Australian Magpie, *Gymnorhina tibcen.* Part 1: Vocal Sharing within and among Social Groups. *Behaviour* 104:1–28.

Brown, H. L., E. K. Urban, and K. Newman. 1982. *The Birds of Africa.* Vol. 1. London: Academic Press.

Brown, J. L. 1974. Alternate Routes to Sociality in Jays—with a Theory for the Evolution of Altruism and Communal Breeding. *American Zoologist* 14:63–80.

———. 1978. Avian Communal Breeding Systems. *Annual Review of Ecology and Systematics* 9:123–56.

Brown, J. L., and E. R. Brown. 1981. Extended Family System in a Communal Bird. *Science* 211:959–60.

Brown, J. L., and J. Kikkawa, eds. 1987. *Animal Societies: Theories and Facts.* Tokyo: Japan Scientific Societies Press.

Brown, L., and D. Amadon. 1968. *Eagles, Hawks and Falcons of the World.* Vol. 1. New York: McGraw-Hill.

Brown, R. E. 1979. Mammalian Social Odours: A Critical Review. In *Advances in the Study of Behaviour* 10, ed. A. Hinde, C. Beer, and M.-C. Busnel, 103–62. New York: Academic Press.

Brown, R. E., and D. W. MacDonald, eds. 1985a. *Social Odours in Mammals.* Vol. 1. Oxford: Clarendon Press.

———. 1985b. *Social Odours in Mammals.* Vol. 2. Oxford: Clarendon Press.

Burghardt, G., and A. S. Rand. 1982. *Iguanas of the World.* Park Ridge, N.J.: Noyes Publishers.

Burley, N. 1981. Evolution of Sexual Indistinguishability. In *Natural Selection and Social Behavior,* ed. R. D. Alexander and D. W. Tinkle, 121–37. New York and Concord: Chiron Press.

Buskirk, W. H. 1976. Social Systems in a Tropical Forest Avifauna. *American Naturalist* 110:293–310.

Buskirk, W. H., G. V. N. Powell, J. F. Wittenberger, R. E. Buskirk, and T. Powell. 1972. Interspecific Bird Flocks in Tropical Highland Panama. *Auk* 89:612–24.

Cairns, R. B. 1986. An Evolutionary and Developmental Perspective on Aggressive Patterns. In *Altruism and Aggression,* ed. C. Zahn-Waxler, E. M. Cummings, and R. Iannoti, 58–87. Cambridge: Cambridge University Press.

Caro, T. M. 1994. *Cheetahs of the Serengeti Plains: Group Living in an Asocial Species.* Chicago: University of Chicago Press.

Carothers, J. H., and F. M. Jacsic. 1984. Time as a Niche Factor: The Role of Interference Competition. *Oikos* 42:403–6.

Carpenter, C. C., and G. W. Ferguson. 1977. Variation and Evolution of

Stereotyped Behavior in Reptiles. Part 1. In *Biology of the Reptilia* 7:335–403. London: Academic Press.

Carpenter, C. R. 1934. A Field Study of the Behavior and Social Relations of Howling Monkeys. *Comparative Psychological Monographs* 10:1–168.

Carpenter, F. L. 1976. Ecology and Evolution of an Andean Hummingbird, *Oreotrochilus estella. University of California Publications in Zoology* 106:1–75.

Carthy, J. D., and F. J. Ebling. 1964. *The Natural History of Aggression.* London and New York: Academic Press.

Caryl, P. 1979. Communication by Agonistic Displays: What Can Games Theory Contribute to Ethology? *Behaviour* 68:136–69.

Case, T. J., and M. E. Gilpin. 1974. Interference Competition and Niche Theory. *Proceedings of the National Academy of Sciences of the United States of America* 71:3073–77.

Catchpole, C., and P. J. B. Slater. 1995. *Bird Song: Biological Themes and Variations.* Cambridge: Cambridge University Press.

Charles-Dominique, P., H. M. Cooper, A. Hladik, C. M. Hladik, C. M. Pages, G. F. Pariente, A. Petter-Rousseau, A. Schilling, and J.-J. Petter. 1980. *Nocturnal Malagasy Primates: Ecology, Physiology and Behavior.* New York: Academic Press.

Cheney, D. L., and R. M. Seyfarth. 1989. Redirected Aggression and Reconciliation among Vervet Monkeys, *Cercopithecus aethiops. Behaviour* 110:258–75.

Chivers, J. J., ed. 1980. *Malayan Forest Primates: Ten Years' Study in Tropical Rain Forest.* New York and London: Plenum Press.

Cifelli, R. F. 1985. South American Ungulate Evolution and Extinction. In *The Great American Biotic Interchange,* ed. F. G. Stehli and S. D. Webb, 249–66. New York and London: Plenum Press.

Clutton-Brock, T. H., and G. A. Parker. 1995a. Punishment in Animal Societies. *Nature* 373 (6511): 209–16.

———. 1995b. Sexual Coercion in Animal Societies. *Animal Behaviour* 49:1345–65.

Coates, B. J. 1985. *The Birds of Papua New Guinea, Including the Bismarck Archipelago and Bougainville.* Vol. 1, *Nonpasserines.* Alderby: Dove.

Colman, A., ed. 1982. *Cooperation and Competition.* U.K.: Van Nostrand and Reinhold.

Colwell, K. 1973. Competition and Coexistence in a Simple Tropical Community. *American Naturalist* 107:737–60.

Connell, J. H. 1983. On the Prevalence and Relative Importance of Interspecific Competition: Evidence from Field Experiments. *American Naturalist* 122 (5): 661–96.

Connor, R. S. 1995. Altruism among Non-relatives: Alternatives to the "Prisoner's Dilemma." *Trends in Ecology & Evolution* 10:84–86.

Cords, M. 1988. Resolution of Aggressive Conflicts by Immature Long-tailed Macaques, *Macaca fascicularis. Animal Behaviour* 36:1124–35.

Cowlishaw, G. 1992. Song Function in Gibbons. *Behaviour* 121: 131–53.

Cracraft, J. 1990. The Origin of Evolutionary Novelties: Patterns and Process at Different Hierarchical Levels. In *Evolutionary Innovations,* ed. M. H. Nitecki, 21–44. Chicago: University of Chicago Press.
Crawford, C. B. 1993. The Future of Sociobiology: Counting Babies or Studying Proximate Mechanisms. *Trends in Ecology & Evolution* 8 (5): 183–86.
Cristol. D. A. 1995. The Coat-Tail Effect in Merged Flocks of Dark-Eyed Juncos: Social Status Depends upon Familiarity. *Animal Behaviour* 50:151–59.
Cronin, H. 1991. *The Ant and the Peacock.* Cambridge: Cambridge University Press.
Culler, J. 1982. *On Deconstruction: Theory and Criticism after Structuralism.* Ithaca: Cornell University Press.
Curio, E. 1994. *Causal and Functional Questions: How Are They Linked? Animal Behaviour* 47:999–1021.
Curry-Lindahl, K. 1981. *Bird Migration in Africa.* Vol. 1. London: Academic Press.
Daanje, A. 1950. On Locomotory Movements in Birds and the Intention Movements Derived from Them. *Behaviour* 3:48–98.
Davies, N. B. 1982. Behaviour and Competition for Scarce Resources. In *Current Problems in Sociobiology,* ed. King's College Sociobiology Group, 363–80. Cambridge: Cambridge University Press.
———. 1989. Sexual Conflict and the Polygamy Threshold. *Animal Behaviour* 38:226–34.
———. 1992. *Dunnock Behaviour and Social Evolution.* Oxford: Oxford University Press.
Davis, W. J. 1985. Acoustic Signaling in the Belted Kingfisher, *Ceryle alcyon.* Ph.D. diss., University of Texas, Austin.
Dawkins, R. 1976. *The Selfish Gene.* Oxford: Oxford University Press.
———. 1982. Replicators and Vehicles. In *Current Problems in Sociobiology,* ed. King's College Sociobiology Group, 45–64. Cambridge: Cambridge University Press.
———. 1996. *Climbing Mount Improbable.* New York: W. W. Norton.
Dawkins, R., and J. R. Krebs. 1978. Animal Signs: Information or Manipulation. In *Behavioural Ecology: An Evolutionary Approach,* ed. R. Krebs and N. B. Davies, 282–309. Oxford: Blackwell.
Delacour, J. 1951. *The Pheasants of the World.* London: Country Life.
Dewsbury, D. A. 1992. On the Problems Studied in Ethology, Comparative Psychology, and Animal Behavior. *Ethology* 92:89–107.
Dimarco, P. L., and R. T. Hanlon. Agonistic Behavior in the Squid *Loligo plei:* Fighting Tactics and the Effects of Resource Value. *Ethology,* in press.
Douthwaite, R. J. 1973. Pied Kingfisher *Ceryle rudis* Populations. *Ostrich* 44:89–94.
———. 1978. Breeding Biology of the Pied Kingfisher *Ceryle rudis* on Lake Victoria. *Journal of the East African Natural History Society and National Museum* 166:1–12.

DuMond, F. V. 1968. The Squirrel Monkey in a Seminatural Environment. In *The Squirrel Monkey*, ed. L. A. Rosenblum and R. W. Cooper, 87–145. New York and London: Academic Press.

Dunbar, R. I. M. 1995a. The Mating System of Callitrichid Primates. 1. Conditions for the Coevolution of Pair Bonding and Twinning. *Animal Behaviour* 50:1057–70.

———. 1995b. The Mating System of Callitrichid Primates. 2. The Impact of Helpers. *Animal Behaviour* 50:1071–89.

Dunham, D. W. 1966. Agonistic Behaviour in Captive Rosebreasted Grosbeaks *Pheucticus ludovicianus* (L.). *Behaviour* 27:160–73.

Du Plessis, M. A. 1991. The Role of Helpers in Feeding Chicks in Cooperatively Breeding Green (Red-Billed) Woodhoopoes. *Behavioral Ecology and Sociobiology* 28:291–95.

———. 1993. Helping Behaviour in Cooperatively Breeding Green Woodhoopoes: Selected or Unselected Trait? *Behaviour* 1–2:49–65.

Eason, P., and S. J. Hannon. 1994. New Birds on the Block: New Neighbors Increase Defensive Costs for Territorial Male Willow Ptarmigan. *Behavioral Ecology and Sociobiology* 34:419–26.

East, M. L., H. Hofer, and W. Wickler. 1993. The Erect "Penis" is a Flag of Submission in a Female-Dominated Society: Greetings in Serengeti Spotted Hyenas. *Behavioral Ecology and Sociobiology* 33:355–70.

Eberhard, W. G. 1985. *Sexual Selection and Animal Genitalia.* Cambridge: Harvard University Press.

Eberhardt, L. S. 1994. Oxygen Consumption during Singing by Male Carolina Wrens *(Thryothorus ludovicianus). Auk* 111:124–30.

Eco, V. 1975. *Trattato di semiotica generale.* Milano: Bompiani.

———. 1988. *Sémiotique et philosophie du langage.* Paris: Presse Universitaire France.

Eisenberg, J. F. 1981. *The Mammalian Radiations.* Chicago: University of Chicago Press.

Eisenberg, J. F., and D. Kleiman. 1972. Olfactory Communication in Mammals. *Annual Review of Ecology and Systematics* 3:1–32.

Emlen, S. T. 1981. Altruism, Kinship and Reciprocity in the White-Fronted Bee-Eater. In *Natural Selection and Social Behavior,* ed. R. D. Alexander and D. Tinkle, 217–30. New York: Chiron Press.

———. 1982a. The Evolution of Helping. 1. An Ecological Constraints Model. *American Naturalist* 119:29–39.

———. 1982b. The Evolution of Helping. 2. The Role of Behavioral Conflict. *American Naturalist* 119:40–53.

———. 1990. The White-Fronted Bee-Eater: Helping in a Colonially Nesting Species. In *Cooperative Breeding in Birds: Longterm Studies of Ecology and Behaviour,* ed. P. B. Stacey and W. K. Koenig, 305–39. Cambridge: Cambridge University Press.

Emlen, S. T., and. N. J. Demong. 1980. Bee-Eaters: An Alternative Route to

Cooperative Breeding. In *Acta 17 Congressus Internationalis Ornithologici* 2:895-901. Berlin: Verlag der Deutschen Ornithologen-Gesellschaft.

———. 1984. Bee-eaters of Baharini. *Natural History* 84(10):50-59.

Emlen, S. T., N. J. Demong, and D. J. Emlen. 1989. Experimental Induction of Infanticide in Female Wattled Jacanas. *Auk* 106:1-7.

Emlen, S. T., and S. L. Vehrencamp. 1983. Cooperative Breeding Strategies among Birds. In *Perspectives in Ornithology,* ed. A. H. Brush and G. A. Clark Jr., 93-120. Cambridge: Cambridge University Press.

Emlen, S. T., and P. H. Wrege. 1991. Breeding Biology of White-Fronted Bee-Eaters at Nakuru: The Influence of Helpers on Breeder Fitness. *Journal of Animal Ecology* 60:309-26.

———. 1992. Parent-Offspring Conflict and the Recruitment of Helpers among Bee-Eaters. *Nature* 356:331-33.

Enquist, M. 1985. Communication during Aggressive Interactions with Particular Reference to Variation in Choice of Behavior. *Animal Behaviour* 33:1152-61.

Enquist, M., and O. Leimar. 1990. The Evolution of Fatal Fighting. *Animal Behaviour* 39:1-9.

———. 1993. The Evolution of Cooperation in Mobile Organisms. *Animal Behaviour* 45:747-57.

Epple, G., and Y. Katz. 1983. The Saddle Back Tamarin and Other Tamarins. In *Reproduction in New World Primates: New Models in Medical Science,* ed. J. Hearn, 115-48. Lancaster, Boston, The Hague: MTP Press.

Ewer, R. F. 1968. *Ethology of Mammals.* New York: Plenum Press.

Feduccia, A. 1995. Explosive Evolution in Tertiary Birds and Mammals. *Science* 267:637-38.

Ferkin, M. H., E. S. Sorokin, and R. E. Johnston. 1995. Seasonal Changes in Scents and Responses to Them in Meadow Voles: Evidence for the Co-evolution of Signals and Response Mechanisms. *Ethology* 180:89-99.

Fisher, J. 1954. Evolution and Bird Sociality. In *Evolution as a Process,* ed. J. Huxley, A. C. Hardy, and E. B. Ford, 71-83. London: George Allen and Unwin Press.

Ford, H. A., and D. C. Paton. 1985. Habitat Selection in Australian Honey Eaters, with Special Reference to Nectar Productivity. In *Habitat Selection in Birds,* ed. M. L. Cody, 367-88. San Diego and New York: Academic Press.

Forshaw, J. M. 1983. *Kingfishers and Related Birds.* Vol. 1, *Alcedinidae: Ceryle to Cittura.* Melbourne: Lansdowne Editions.

———. 1985. *Kingfishers and Related Birds.* Vol. 2, *Alcedinidae: Halcyon to Tanysiptera.* Melbourne: Lansdowne Editions.

———. 1987. *Kingfishers and Related Birds.* Vol. 3, *Todidae, Momotidae, Meropidae.* Melbourne: Lansdowne Editions.

———. 1991. *Kingfishers and Related Birds.* Vol. 4, *Leptosomatidae, Coraciidae, Upupidae, Phoeniculidae.* Melbourne: Lansdowne Editions.

Frank, L. G., S. E. Glickman, and P. Licht. 1991. Fatal Sibling Aggression, Precocial Development and Androgens in Neonatal Spotted Hyenas. *Science* 252:702–4.

Frank, L. G., M. L. Weldele, and S. E. Glickman. 1995. Masculinization in Hyaenas. *Nature* 377:584–85.

Frazzetta, T. H. 1975. *Complex Adaptations in Evolving Populations.* Sunderland, Mass.: Sinauer Associates.

Fry, C. H. 1980. The Evolutionary Biology of Kingfishers (Alcedinidae). *Living Bird* 18:113–60.

———. 1984. *The Bee-Eaters.* Calton: T. & A. D. Poyser.

Fry, C. H., K. Fry, and A. Harris. 1992. *Kingfishers, Bee-Eaters and Rollers.* Princeton: Princeton University Press.

Fry, C. H., S. Keith, and E. K. Urban. 1988. *The Birds of Africa.* Vol. 3. London: Academic Press.

Fujioka, Masahiro. 1987. Mechanisms and Factors of Brood Reduction in the Cattle Egret *Bubulcus ibis.* In *Animal Societies: Theories and Facts,* ed. J .L. Brown and J. Kikkawa, 115–23. Tokyo: Japan Scientific Societies Press.

Furness, R. W. 1987. *The Skuas.* Calton: T. & A. D. Poyser.

Furuichi, T., and H. Ihobe. 1994. Variation in Male Relationships in Bonobos and Chimpanzees. *Behaviour* 130:211–28.

Gans, C., and D. W. Tinkle, eds. 1977. *The Biology of the Reptilia.* Vol. 7. London: Academic Press.

Gaunt, A. S., T. L. Bucher, S. L. L. Gaunt, and L. F. Baptista. 1996. Is Singing Costly? *Auk* 113:718–21.

Gauthier-Pilters, H., and A. I. Dagg. 1981. *The Camel: Its Evolution, Ecology, Behavior, and Relationship to Man.* Chicago: University of Chicago Press.

Gautier, J. Y., P. Deleporte, and C. Rivault. 1988. Relationships between Ecology and Social Behavior in Cockroaches. In *The Ecology of Social Behavior,* ed. C. N. Slobodnicoff. San Diego: Academic Press.

Gautier-Hion, A. 1988. Polyspecific Associations among Forest Quenons: Ecological, Behavioural and Evolutionary Aspects. In *A Primate Radiation: Evolutionary Biology of the African Quenona,* ed. A. Gautier-Hion, F. Bourlière, and J. P. Gantier, 452–76. New York: Cambridge University Press.

Gautier-Hion, A., F. Bourlière, J. P. Gauthier, and J. Kingdon, eds. 1988. *A Primate Radiation: Evolutionary Biology of the African Guenons.* Cambridge: Cambridge University Press.

Gautier-Pilters, H., and A. I. Dagg. 1981. *The Camel.* Chicago and London: University of Chicago Press.

Godfrey, H. C. J. 1995. Evolutionary Theory of Parent-Offspring Conflict. *Nature* 376 (6536): 133–38.

Goodwin, D. 1976. *Crows of the World.* Ithaca: Cornell University Press.

———. 1983. *Pigeons and Doves of the World.* British Museum (Natural History). Ithaca: Cornell University Press.

Gorman, M. L., and B. J. Trowbridge. 1989. The Role of Odor in the Social Lives of Carnivores. In *Carnivore Behavior, Ecology, and Evolution,* ed. J. L. Gittelman, 57-88. Chapman and Hall.

Gould, S. J. 1995. A Task for Paleobiology at the Threshold of Majority. *Paleobiology* 21:1-14.

Gould, S. J., and R. C. Lewontin. 1979. The Spandrels of San Marco and the Panglossum Paradigm: A Critique of the Adaptationist Programme. In *The Evolution of Adaptation by Natural Selection,* organized by J. F. Maynard Smith and R. Holliday, 581-604. Cambridge: Cambridge University Press.

Gradwohl, J., and R. Greenberg. 1980. The Formation of Antwren Flocks on Barro Colorado Island, Panama. *Auk* 97:385-95.

Greenewalt, 1960. *Hummingbirds.* Garden City, N.Y.: Doubleday.

Gurnell, J. 1987. *The Natural History of Squirrels.* London: Christopher Helm.

Halliday, T. 1980. *Sexual Strategy.* Chicago: University of Chicago Press.

Hamilton, W. D. 1964a. The Genetical Evolution of Social Behaviour. Part 1. *Journal of Theoretical Biology* 7:1-16.

———. 1964b. The Genetical Evolution of Social Behaviour. Part 2. *Journal of Theoretical Biology* 7:17-52.

———. 1972. Altruism and Related Phenomena, Mainly in Social Insects. *Annual Review of Ecology and Systematics* 3:193-232.

Hammerstein, P., and R. F. Hoekstra. 1995. Mutualism on the Move. *Nature* 376 (6536): 121-22.

Hanlon, R. T. 1982. The Functional Organization of Chromatophores and Iridescent Cells in the Body Patterning of *Loligo plei* (Cephalopoda: Myopsida). *Malacologia* 23 (1): 89-119.

Hanlon, R. T., M. J. Smale, and W. H. H. Sauer. 1994. An Ethogram of Body Patterning Behavior in the Squid *Loligo vulgaris reynaudii* on Spawning Grounds in South Africa. *Biological Bulletin of Wood's Hole* 187 (3): 363-72.

Harcourt, A. H. 1987. Cooperation as a Competitive Strategy in Primates and Birds. In *Animal Societies: Theories and Facts,* ed. Y. Ito, J. L. Brown, and J. Kikkawa, 141-57. Tokyo: Japan Scientific Societies Press.

———. 1992. Conditions and Alliances: Are Primates More Complex than Non-Primates? In *Coalitions and Alliances in Humans and Other Animals,* ed. A. H. Harcourt and F. B. M. de Waal, 445-72. Oxford University Press.

Hardy, J. 1961. *Studies in Behavior and Phylogeny of Certain New World Jays (Garrulinae).* University of Kansas Science Bulletin 42.

Hare, J. E., and J. O. Murie. 1996. Ground Squirrel Sociality and the Quest for the Holy Grail: Does Kinship Influence Behavioral Discrimination by Juvenile Colombian Ground Squirrels? *Behavioral Ecology* 7:76-81.

Hausfater, G., and S. Blaffer Hrdy. 1984. *Infanticide: Comparative and Evolutionary Perspectives.* Hawthorne, N.Y.: Aldine.

Heal, J. 1991. Altruism. In *Cooperation and Prosocial Behavior,* ed. Robert A. Hinde and J. Groebel, 159–72. Cambridge: University of Cambridge Press.

Heiligenberg, W. 1991. The Neural Basis of Behavior: A Neuro-Ethological View. *Annual Review of Neuroscience* 14:247–67.

Heinrich, B. 1988. Food Sharing in the Raven *Corvus corax.* In *The Ecology of Social Behavior,* ed. C. N. Slobodnichoff. San Diego: Academic Press.

———. 1989. *Ravens in Winter.* New York: Summit Books.

Heinroth, O. 1911. Beiträge zur Biologie, namentlich Ethologie und Psychologie der Anatiden. Verhandlungen 5. *Internationaler Ornithologen Kongress* 5:589–702.

Heinsohn, R. 1995. Raid of the Red-Eyed Chicknappers. *Natural History* 104 (2): 45–53.

Herring, P. J. 1977. Luminescence in Cephalopods and Fish. In *The Biology of Cephalopods,* ed. M. Nixon and J. B. Messenger, 127–59. Symposium of the Zoological Society of London 38. London: Academic Press.

Hill, D. A. 1994. Affiliative Behaviour between Adult Males of the Genus *Macaca. Behaviour* 130:293–308.

Hill, D. A., and J. A. R. A. M. van Hooff. 1994. Affiliative Relationships between Males in Groups of Nonhuman Primates: A Summary. *Behaviour* 130:143–49.

Hinde, R. A. 1981. Animal Signals: Ethological and Games Theory Approaches Are Not Incompatible. *Animal Behaviour* 29 (2): 535–42.

———, ed. 1983. *Primate Social Relationships.* Oxford: Blackwell.

Hinde, R. A., and J. Groebel, eds. 1991. *Cooperation and Prosocial Behaviour.* Cambridge: Cambridge University Press.

Hiraiwa-Hasegawa, M. 1987. Infanticide in Primates and a Possible Case of Male-Biased Infanticide in Chimpanzees. In *Animal Societies: Theories and Facts,* ed. Y. Ito, J. L. Brown, and J. Kikkawa, 125–39. Tokyo: Japan Scientific Societies Press.

Hockett, C. F. 1960a. Logical Considerations in the Study of Animal Communication. In *Animal Sound and Communication,* ed. W. E. Lanyon and W. N. Tavolga, 392–420. Washington, D.C.: American Institute of Biological Sciences.

———. 1960b. The Origin of Speech. *Scientific American* 203 (3): 89–96.

Hockett, C. F., and S. A. Altmann. 1968. A Note on Design Features. In *Animal Communication,* ed. T. A. Sebeok, 61–72. Bloomington: Indiana University Press.

Holley, A. J. F. 1993. Do Brown Hares Signal to Foxes? *Ethology* 94:21–30.

van Hooff, J. A. R. A. M., and C. P. van Schaik. 1992. Cooperation in Competition: The Ecology of Primate Bonds. In *Coalitions and Alliances in Humans and Other Animals,* ed. H. A. Harcourt and F. B. M. de Waal, 357–89. Oxford: Oxford University Press.

———. 1994. Male Bonds: Affiliative Relationships among Nonhuman Primate Males. *Behaviour* 130:309–37.

Hoogland, J. L. 1995. *The Black-Tailed Prairie Dog: Social Life of a Burrowing Mammal.* Chicago and London: University of Chicago Press.

Horn, A. G., M. L. Leonard, and D. M. Weary. 1995. Oxygen Consumption during Crowing by Roosters: Talk Is Cheap. *Animal Behaviour* 80:1171–75.

Horwich, R. H. 1972. *The Ontogeny of Social Behavior in the Gray Squirrel (Sciurus carolinensis).* Advances in Ethology 8. Berlin and Hamburg: Paul Parey.

Huntingford, F. A., and A. Turner. 1987. *Animal Conflict.* London: Chapman and Hall.

Imanishi, K., and S. A. Altmann, eds. 1965. *Japanese Monkeys. A Collection of Translations.* Alberta: S. A. Altmann, University of Alberta.

Johnsgard, P. A. 1965. *Handbook of Waterfowl Behavior.* Ithaca: Cornell University Press.

———. 1973. *Grouse and Quails of North America.* Lincoln: University of Nebraska Press.

———. 1983. *Cranes of the World.* Bloomington: Indiana University Press.

———. 1986. *The Pheasants of the World.* Oxford, New York, and Tokyo: Oxford University Press.

Johnstone, R. A., and A. Graffen. 1993. Dishonesty and the Handicap Principal. *Animal Behaviour* 46:759–64.

Johnstone, R. A., and K. Norris. 1993. Badges of Status and the Cost of Aggression. *Behavioral Ecology and Sociobiology* 32:127–34.

Jolly, A. 1961. *Lemur Behavior.* Chicago: University of Chicago Press.

———. 1972. *The Evolution of Primate Behavior.* New York: MacMillan.

Kappeler, P. M., and J. V. Ganz-Horn, eds. 1993. *Lemur Social Systems and their Ecological Basis.* New York: Plenum Press.

Karli, P. 1991. *Animal and Human Aggression.* New York: Oxford University Press.

Kavanagh, M. 1983. *A Complete Guide to Monkeys, Apes and Other Primates.* New York: Viking Press.

Keller, L., and H. K. Reeve. 1994. Partitioning of Reproduction in Animal Societies. *Trends in Ecology and Evolution* 9:98–102.

Kemp, A. C. 1976. A Study of the Ecology, Behaviour and Systematics of *Tockus* Hornbills (Aves: Bucerotidae). *Transvaal Museum Memoirs* 20.

———. 1988. The Behavioural Ecology of the Southern Ground Hornbill: Are Competitive Offspring at a Premium? In *Current Topics in Avian Biology,* 267–71. Bonn: Proceedings of the International-100 Deutsche Ornitologen Gesellschaft Meeting.

———. 1995. *The Hornbills: Bucerotiformes.* Oxford: Oxford University Press.

Keppler, A. K. 1977. Comparative Studies of Todies, with Emphasis on the Puerto Rican Tody, *Todus mexicanus.* Publications of the Nuttall Ornithological Club 16. Cambridge, England.

Kilham, L. 1956. Breeding and Other Habits of Casqued Hornbills *(Bycanistis subcylindricus). Smithsonian Miscellaneous Collections* 131 (9): 1–45.

———. 1961. Reproductive Behavior of Red-bellied Woodpeckers. *Wilson Bulletin* 73:237–54.

———. 1966. Reproductive Behavior of Hairy Woodpeckers. 1. Pair Formation and Courtship. *Wilson Bulletin* 78:251–65.

King, B., and E. C. Dickinson. 1975. *Birds of South-East Asia.* London: Harper Collins.

Kinzey, W. G. 1981. The Titi Monkey, Genus *Callicebus.* In *Ecology and Behavior of Neotropical Primates,* ed. A. F. Coimba-Filho and R. A. Mittermeier, 241–75. Rio de Janeiro: Academia Brasileira de Ciencias.

Kirkpatrick, J. W., and G. F. Woolfenden. 1986. Demographic Routes to Cooperative Breeding in Some New World Jays. In *Evolution of Animal Behavior: Paleontological and Field Approaches,* ed. M. H. Nitecki and J. A. Kitchell, 137–60. New York: Oxford University Press.

Knoder, C. E., and Baillie, R. M. 1956. The Ecology and Breeding Biology of the Reeves' Pheasant on Tappan Island, 1955. *Ohio Wildlife Investigations* 7:36–45.

Koenig, W. D. 1981. Reproductive Success, Group Size, and the Evolution of Cooperative Breeding in the Acorn Woodpecker. *American Naturalist* 117:421–43.

Koenig, W. D., and R. L. Mumme. 1987. *Population Ecology of the Cooperatively Breeding Acorn Woodpecker.* Princeton: Princeton University Press.

Kortlandt, A. 1940. Eine Übersicht der Angeborenen Verhaltensweisen des Mitteleuropäischen Kormorans (*Phalacrocorax carbo sinensis* Shaw and Nodder), Ihre Funktion, Ontogenetische Entwicklung und Phylogenetische Herkunft. *Archives Néerlandaises de Zoologie* 4:401–42.

———. 1972. *New Perspectives on Ape and Human Evolution.* Amsterdam: Stichting voor Psychobiologie.

Krebs, J. R. 1973. Social Learning and the Significance of Mixed Species Flocks of Chickadees (*Parus* spp.). *Canadian Journal of Zoology* 51:1275–88.

Krebs, J. R., and N. B. Davies. 1993. *An Introduction to Behavioural Ecology.* Oxford: Blackwell Scientific Publications.

Kruijt, J. P. 1964. Ontogeny of Social Behaviour in Burmese Jungle Fowl *(Gallus gallus spadiceus). Behaviour* 12:1–201.

Kruuk, H. 1972. *The Spotted Hyena: A Study of Predation and Social Behavior.* Chicago: Chicago University Press.

———. 1989. *The Social Badger: Ecology and Behaviour of a Group-Living Carnivore (Meles meles).* Oxford: Oxford University Press.

Kummer, H. 1968. *Social Organization of Hamadryas Baboons: A Field Study.* Bibliotica 6. Basel, N.Y.: S. Karger.

Lack, D. 1968. *Ecological Adaptations for Breeding in Birds.* London: Methuen.

Lande, R. 1980. Sexual Dimorphism, Sexual Selection, and Adaptation in Polygenic Characters. *Evolution* 292–305.

van Lawick, H., and J. van Lawick-Goodall. 1970. *The Innocent Killers.* London: Collins.

van Lawick-Goodall, J., and H. van Lawick. 1971. *In the Shadow of Man.* Boston: Houghton Mifflin.

Lazarus, J. 1982. Competition and Conflict in Animals. In *Cooperation and Competition in Animals and Humans,* ed. A. M. Colman, 26–56. Wokingham, England: Van Nostrand and Reinhall.

Le Boeuf, B. 1985. *Elephant Seals.* Pacific Grove, Calif.: Boxwood Press.

Lefebvre, L., B. Palameta, and K. E. Hatch. 1996. Is Group Living Associated with Social Learning? A Comparative Test of a Gregarious and a Territorial Columbid. *Behaviour* 1331:241–61.

Leigh, E. G., Jr. The Evolution of Mutualism and Other Forms of Harmony at Various Levels of Biological Organization. *Écologie,* in press.

Leighton, M. 1982. Fruit Resource and Patterns of Feeding, Spacing, and Grouping among Sympatric Bornean Hornbills (Bucerotidae). Ph.D. diss., University of California, Davis.

———. 1986. Hornbill Social Dispersion: Variation on a Monogamous Theme. In *Ecological Aspects of Social Evolution: Birds and Mammals,* ed. D. J. Rubenstein and R. W. Wrangham, 108–30. Princeton: Princeton University Press.

Leighton, M., and D. R. Leighton. 1983. Vertebrate Responses to Fruiting Seasonality within a Bornean Rain Forest. In *Tropical Rain Forest: Ecology and Management,* ed. S. L. Sutton, T. C. Whitmore, and A. C. Chandwick, 181–96. Oxford: Blackwell.

Leyhausen, P. 1956. Verhaltensstudien an Katzen. Supplement 2. *Zeitschrift für Tierpsychologie.*

Ligon, J. D. 1983. Commentary. In *Perspectives in Ornithology,* ed. A. H. Brush and G. A. Clark Jr., 120–27. Cambridge: Cambridge University Press.

Ligon, J. D., and S. H. Ligon. 1978a. Communal Breeding in Green Woodhoopoes as a Case for Reciprocity. *Nature* 276:496–98.

———. 1978b. The Communal Social System of the Green Woodhoopoe in Kenya. *Living Bird* 17:159–97.

———. 1981. Demographic Patterns and Cooperative Breeding in the Green Woodhoopoe *Phoeniculus purpureus.* In *Natural Selection and Social Behaviour: Recent Research and New Theory,* ed. R. D. Alexander and D. W. Tinkle, 231–43. New York: Chiron Press.

———. 1983. Reciprocity in Green Woodhoopoes *Phoeniculus purpureus. Animal Behaviour* 31:480–89.

———. 1990. Green Woodhoopoes: Life History Traits and Sociality. In *Cooperative Breeding in Birds,* ed. P. B. Stacey and W. D. Koenig, 33–65. Cambridge: Cambridge University Press.

Ligon, J. D., and P. B. Stacey. 1989. On the Significance of Helping Behavior in Birds. *Auk* 106:700–5.

———. 1991. The Origin and Maintenance of Helping Behavior in Birds. *American Naturalist* 138:254–58.

Lillegraven, J. A., Z. Kielan-Jaworowska, and W. A. Clemens. 1979. *Mesozoic Mammals: The First Two-Thirds of Mammalian History.* Berkeley: University of California Press.

Lindell, C. 1996. Benefits and Costs to Plain-Fronted Thornbirds *(Phacellodomus rufifrons)* of Interactions with Avian Nest Associates. *Auk* 113:565–77.

Lloyd, J. E. 1983. Bioluminescence and Communication in Insects. *Annual Review of Entomology* 28:131–60.

Lorenz, K. 1931. Beiträge zur Ethologie Sozialer Corviden. *Journal für Ornithologie* 79:67–127.

———. 1935. Der Kumpane in der Umwelt des Vogels. *Journal für Ornithologie* 83:137–215, 289–413.

———. 1939. Vergleichende Verhaltensforschung. Supplement, *Deutscher Zoologischer Anzeiger* 12:69–102.

———. 1941. Vergleichende Bewegungsstudien an Anatinen. Supplement, *Journal für Ornithologie* 89:194–294.

———. 1951–53. Comparative Studies on the Behaviour of the Anatinae. *Avicultural Magazine* 57:157–82; 58:8–17, 61–72, 86–94, 172–84; 59:24–34, 80–91.

———. 1952. *King Solomon's Ring.* New York: Crowell.

———. 1963. *Das Sogenannte Böse.* Wien: Borotha-Schoeler Verlag.

Lyon, D. L., and C. Chader. 1971. Exploitation of Nectar Resources by Hummingbirds, Bees *(Bombus)* and *Diglossa baritula* and Its Role in Evolution of *Penstemon kunthii. Condor* 73:246–48.

MacDonald, D. W. 1985. The Carnivores: Order Carnivora. In *Social Odours in Mammals,* ed. R. E. Brown and W. D. MacDonald, 619–722. Oxford: Clarendon Press.

MacKinnon, J. R. 1974. The Behaviour and Ecology of Wild Orang-Utans *(Pongo pygmaeus). Animal Behaviour* 22:3–74.

MacLean, P. D. 1964. Mirror Disputes in the Squirrel Monkey, *Saimiri sciureus. Science* 146:950–52.

MacRoberts, M. H., and B. R. MacRoberts. 1976. *Social Organization and Behavior of the Acorn Woodpecker in Central Coastal California.* Ornithological Monographs 21. Washington, D.C.: American Ornithologists' Union.

Marshall, J. T., and E. R. Marshall. 1976. Gibbons and Their Territorial Songs. *Science* 193:235–37.

Martin, G. 1996. Birds in Double Trouble. *Nature* 380 (6376): 666–67.

Martins, E. P. 1994. Structural Complexity in a Lizard Communication System: The *Sceloporus graciosus* Push-Up Display. *Copeia* 1994 (4): 944–55.

Marzluff, J. M., and R. B. Balda. 1988. Resource and Climatic Variability: Influences on Sociality of Two Southwestern Corvids. In *The Ecology of Social Behavior,* ed. C. N. Slobodchikoff, 255–83. San Diego: Academic Press.

Mason, W. A. 1966. Social Organization of the South American Monkey, *Callicebus moloch: A Preliminary Report. Tulane Studies in Zoology* 13:23–28.

Mason, W. A., and S. P. Mendoza, eds. 1993. *Primate Social Conflict.* Albany: New York Press.

Mather, J. A., and D. L. Mather. 1994. Skin Colours and Patterns of Juvenile *Octopus vulgaris* (Mollusca Cephalopoda) in Bermuda. *Vie et Milieu* 44:267–72.

Maurer, E. A. 1984. Interference and Exploitation in Bird Communication. *Wilson Bulletin* 96:380–95.

Maynard Smith, J. 1974. The Theory of Games and the Evolution of Animal Conflicts. *Journal of Theoretical Biology* 47:20–221.

———. 1979. Game Theory and the Evolution of Behaviour. *Proceedings of the Royal Society of London B Biological Sciences* 205 (1161): 475–88.

———. 1982a. *Evolution and the Theory of Games.* Cambridge: Cambridge University Press.

———. 1982b. Do Animals Convey Information about Their Intentions? *Journal of Theoretical Biology* 97 (1): 1–5.

———. 1984. Game Theory and the Evolution of Behaviour. *The Behavioral and Brain Sciences* 7:95–125.

———. 1991. Honest Signalling: The Philip Sydney Game. *Animal Behaviour* 42:1034–1135.

———. 1994. Must Reliable Signals Always Be Costly? *Animal Behaviour* 47:1115–20.

Maynard Smith, J., and J. Harper. 1988. The Evolution of Aggression: Can Selection Generate Variability? *Philosophical Transactions of the Royal Society of London B* 319:557–70.

Maynard Smith, J., and M. G. Ridpath. 1972. *Wife Sharing in the Tasmanian Native Hen, Tribonyx mortierii: A Case of Kin Selection? American Naturalist* 196:447–52.

Mayr, E. 1982. *The Growth of Biological Thought.* Cambridge, Mass., and London: Belknap Press.

———. 1983. How to Carry out the Adaptationist Program? *American Naturalist* 121 (3): 322–34.

McGrew, W. C. 1992. *Chimpanzee Material Culture.* Cambridge: Cambridge University Press.

McKenna, J. 1978. Biosocial Functions of Grooming Behaviour among the Common Indian Langur Monkey *(Presbytis entellus). American Journal of Physical Anthropology* 48 (4): 503–9.

McKinney, D. G. 1961. Analysis of the Displays of the European Eider *Somateria mollissima mollissima* (L.) and the Pacific Eider *Somateria mollissima v. nigra* Bonaparte. Supplement, *Behaviour* 7:1–123.

Menzel, C. R. 1993. Coordination and Conflict in Callicebus Social Groups. In *Primate Social Conflict,* ed. W. A. Mason and S. P. Mendoza, 253–90. New York: State University of New York Press.

Merila, J., and D. A. Wiggins. 1995. Interspecific Competition for Nest Holes Causes Adult Mortality in the Collared Flycatcher. *Condor* 97:445–50.

Miller, R. J. 1978. Agonistic Behavior in Fishes and Terrestrial Vertebrates. In *Contrasts in Behavior,* ed. E. S. Reese and F. J. Lighter, 281–310. New York: John Wiley and Sons.

Miller, R. S. 1968. Conditions of Competition between Redwings and Yellowheaded Blackbirds. *Journal of Animal Ecology* 37:43–62.

Mills, M. G .L. 1989. The Comparative Ecology of Hyenas: The Importance of Diet and Food Dispersion. In *Carnivore Behavior, Ecology and Evolution,* ed. J. L. Gittelman, 125–68. London: Chapman and Hall.

Mills, M. G. L., and M. L. Gorman. 1987. The Scent Marking Behavior of the Spotted Hyaena, *Crocuta crocuta,* in the Southern Kalahari. *Journal of Zoology* (Zoological Society of London) 212:483–97.

Mills, M. G. L., M. L. Gorman, and M. E. J. Mills. 1980. The Scent Marking Behavior of the Brown Hyaena, *Hyaena brunnea,* in the Southern Kalahari. *South African Journal of Zoology* 15:240–48.

Milon, P., J.-J. Petter, and G. Randrianasolo. 1973. *Faune de Madagascar.* Oiseaux 35. Organisation de Recherche Scientifique d'Outre-Mer, Tananarive. Paris: Centre Nacional de la Recherche Scientifique.

Minot, E. O. 1981. Effects of Interspecific Competition for Food in Breeding Blue and Great Tits. *Journal of Animal Ecology* 50:375–85.

Minot, E. O., and C. M. Perrins. 1986. Interspecific Interference Competition: Nest Sites for Blue and Great Tits. *Journal of Animal Ecology* 55:331–50.

Mitchell, C. L. 1994. Migration Alliances and Coalitions among Adult Male South American Squirrel Monkeys *(Saimiri sciureus). Behaviour* 130:169–90.

Mock, D. W. 1984a. Infanticide Siblicide, and Avian Nesting Mortality. In *Infanticide: Comparative and Evolutionary Perspectives,* ed. G. Hausvater and S. Blaffer Hrdy, 2–30. New York: Aldine.

———. 1984b. Siblicidal Aggression and Resource Monopolization in Birds. *Science* 731–33.

Møller, A. P. 1994. *Sexual Selection and the Barn Swallow.* Oxford Series in Ecology and Evolution. Oxford and New York: Oxford University Press.

Montgomery, G. G., and M. E. Sunquist. 1975. Impact of Sloths on Neotropical Forest, Energy Flow and Nutrient Cycling. In *Tropical Ecological Systems: Trends in Terrestrial and Aquatic Research,* ed. F. B. Golley and E. Medina, 69–98. New York: Springer Verlag.

———. 1978. Habitat Selection and Use by Two-Toed and Three-Toed Sloths. In *The Ecology of Arborial Folivores,* ed. G. G. Montgomery, 329–59. Washington, D.C.: Smithsonian Institution Press.

Mooring, M. S., and B. L. Hart. 1995. Costs of Allogrooming in Impala: Distraction from Vigilance. *Animal Behaviour* 49:1414–16.

Morel, G. J., and M. Y. Morel. 1982. Dates de reproduction des oiseaux de la Sénégambie. *Bonner Zoologische Beiträge* 33:249–68.

Morris, R., and D. Morris. 1966. *Men and Apes.* London: Hutchinson.

Morton, E. S. 1977. On the Occurrence and Significance of Motivation-Structural Rules in Some Birds and Mammal Sounds. *American Naturalist* 111:855–69.

———. 1982. Grading, Discreteness, Redundance, and Motivation-Structural Rules. In *Acoustic Communication in Birds.* vol. 1, ed. D. E. Kroodsma, E. H. Miller, and H. Ouellet, 183–212. New York: Academic Press.

———. 1986. Predictions from the Ranging Hypothesis for the Evolution of Long Distance Signals in Birds. *Behaviour* 99:65–86.

———. 1994. Sound Symbolism and Its Role in Non-Human Vertebrate Communication. In *Sound Symbolim,* ed. L. Hinton, J. Nichols, and J. J. Ohala, 348–65. Cambridge: Cambridge University Press.

Morton, E. S., and J. Page. 1992. *Animal Talk: Science and the Voices of Nature.* New York: Random House.

Moynihan, M. 1962a. Hostile and Sexual Behavior Patterns of South American and Pacific Laridae. Supplement, *Behaviour* 7:1–365.

———. 1962b. The Organization and Probable Evolution of Some Mixed Species Flocks of Neotropical Birds. *Smithsonian Miscellaneous Collections* 149 (5): 1–34.

———. 1966. Communication in the Titi Monkey Callicebus. *Journal of Zoology* (Zoological Society of London) 150:77–127.

———. 1968. Social Mimicry: Character Convergence vs. Character Displacement. *Evolution* 22:315–31.

———. 1970a. Some Behavior Patterns of Platyrrhine Monkeys. 2. *Saguinus geoffroyi* and Some Other Tamarins. *Smithsonian Contributions to Zoology* 28:1–77.

———. 1970b. Control, Suppression, Decay, Disappearance and Replacement of Displays. *Journal of Theoretical Biology* 29:85–112.

———. 1976. *The New World Primates: Adaptive Radiation and the Evolution of Social Behavior, Languages, and Intelligence.* Princeton: Princeton University Press.

———. 1978. An "Ad Hoc" Association of Hornbills, Starlings, Coucals and Other Birds. *Terre et la Vie* 32:357–76.

———. 1979. *Geographic Variation in Social Behavior and in Adaptations to Competition among Andean Birds.* Ed. Raymond A. Paynter Jr. Publications of the Nuttall Ornithological Club 18. Cambridge, England.

———. 1982a. Why Is Lying about Intentions Rare during Some Kinds of Contests? *Journal of Theoretical Biology* 97 (1): 7–12.

———. 1982b. Spatial Parameters and Interspecific Social Relations: Some Differences between Birds and Fishes. *Journal of Theoretical Biology* 95:253–62.

———. 1983a. Notes on the Behavior of *Euprymna scolopes* (Cephalopoda: Sepiolidae). *Behaviour* 85:25–41.

———. 1983b. Notes on the Behavior of *Idiosepius pygmaeus* (Cephalopoda: Idiosepiidae). *Behaviour* 85:42–57.

———. 1985. *Communication and Non-Communication by Cephalopods.* Bloomington: Indiana University Press.

———. 1987a. Social Relations among Halcyon Kingfishers in Sénégal. *Terre et la Vie* 42:145–66.

———. 1987b. Notes on the Behavior of Giant Kingfishers. *Malimbus* 9:97–104.

———. 1988. The Opportunism of the Abyssinian Roller *(Coracias abyssinica)* in Sénégal. *Terre et la Vie* 43:159–66.

———. 1990. Social, Sexual and Pseudosexual Behavior of the Blue-Bellied Roller, *Coracias cyanogaster:* The Consequences of Crowding or Concentration. *Smithsonian Contributions to Zoology* 491:1–21.

———. 1991. Structures of Animal Communication. In *Man and Beast Revisited,* ed. M. H. Robinson and L. Tiger, 193–202. Washington, D.C.: Smithsonian Institution Press.

———. 1995. Social Structures and Behavior Patterns of Captive and Feral Reeves' Pheasants *(Syrmaticus reevesi)* in France. *Alauda* 63:221–28.

———. 1996. Self Awareness in Coleoid Cephalopods. In *Anthropomorphism Anecdotes and Animals,* ed. R. W. Mitchell, N. S. Thompson, and H. L. Miles. Binghamton: State University of New York Press.

———. *On the Various Kinds of Infanticide.* In press.

Moynihan, M., and A. F. Rodaniche. 1977. The Behavior and Natural History of the Caribbean Reef Squid *Sepioteuthis sepioidea. Advances in Ethology* 25:1–150.

Murphy, R. C. 1936a. *Oceanic Birds of South America.* Vol. 1. New York: American Museum of Natural History.

———. 1936b. *Oceanic Birds of South America.* Vol. 2. New York: American Museum of Natural History.

Neelakantan, K. K. 1962. Drumming by, and an Instance of Homosexual Behavior in, the Lesser Golden-backed Woodpecker *(Dinopium benghalense). Journal of the Bombay Natural History Society* 59:288–90.

Nelson, J. B. 1978. *The Sulidae: Gannets and Boobies.* Oxford: Oxford University Press.

Nevo, E., S. Simson, G. Heth, and A. Beiles. 1992. Adaptive Pacifistic Behaviour in Subterraneum Male Rats in the Sahara Desert, Contrasting to and Originating from Polymorphic Aggression in Israeli species. *Behaviour* 123:70–76.

Oniki, Y. 1985. Why Robin Eggs Are Blue and Birds Build Nests: Statistical Tests for Amazonian Birds. In *Neotropical Ornithology,* 536–45. Ornithological Monographs 36. Washington, D.C.: American Ornithologists' Union.

O'Riain, M. J., J. U. M. Jarves, and C. G. Faulkes. 1996. A Dispersive Morph in the Naked Mole Rat. *Nature* 380 (6675): 619–21.

Packard, A. 1988a. The Skin of Cephalopods (Coleoids: General and Special Adaptations). In *The Mollusca,* vol. 2, *Form and Function,* ed. E. R. Trueman and M. R. Clarke, 37–67. New York: Academic Press.

———. 1988b. Visual Tactics and Evolutionary Strategies. In *Cephalopods: Present and Past,* ed. J. Wiedmann and J. Kullmann, 89–103. Stuttgart: Schweizbart'sche Verlagsbuchhandlung.

Packard, A., and G. D. Sanders. 1971. Body-Patterns of *Octopus vulgaris* and Maturation of the Response to Disturbance. *Animal Behaviour* 19:780–90.

Packer, C. 1986a. Whatever Happened to Reciprocal Altruism? *Trends in Ecology & Evolution* 1:142–43.

———. 1986b. The Ecology of Sociality in Felids. In *Ecological Aspects of Social Evolution,* ed. D. J. Rubenstein and R. W. Wrangham, 429–51. Princeton: Princeton University Press.

Packer, C., and A. E. Pusey. 1983a. Male Takeovers and Female Reproductive Parameters: A Simulation of Oestrus Synchrony in Lions *(Panthera leo). Animal Behaviour* 31:334–40.

———. 1983b. Adaptations of Female Lions to Infanticide by Incoming Males. *American Naturalist* 121:716–28.

Parker, P. G., T. A. Waite, and D. M. Decker. 1995. Kinship and Association in Communally Roosting Black Vultures. *Animal Behaviour* 49:395–401.

Parmigiani, S., and F. S. vom Saal, eds. 1994. *Infanticide and Parental Care.* Switzerland: Harwood Academic Publishers.

Payne, R. S., and S. McVay. 1971. Songs of Humpback Whales. *Science* 173:585–97.

Peeke, H. V. S., and L. Petrinovich, eds. 1983. *Habituation, Sensitization, and Behavior.* New York: Academic Press.

Peres, C. A. 1992. Consequences of Joint Territoriality in a Mixed Species Group of Tamarin Monkeys. *Behaviour* 123:220–46.

Petrie, M. 1992. Copulation Frequency in Birds: Why Do Females Copulate More than Once with the Same Male? *Animal Behaviour* 44:790–92.

Petter, J.-J., R. Albignac, and Y. Rumpler. 1977. *Mammifères lémuriens (primates prosimiens).* Faune de Madagascar 44. Paris: Institut Français de Recherche Scientifique pour le Développement Coopération, France (ORSTOM) and Centre Nacional de la Recherche Scientifique (CNRS).

Pfennig, D. W., and P. W. Sherman. 1995. Kin Recognition. *Scientific American* 272 (6): 98–103.

Pierce, G. W. 1948. *The Songs of Insects.* Cambridge: Harvard University Press.

Pitcher, T. J., ed. 1986. *The Behavior of Teleost Fishes.* Baltimore: Johns Hopkins University Press.

———. 1993. *Behaviour of Teleost Fishes.* London: Chapman and Hall.

Ploog, D. W. 1967. The Behavior of Squirrel Monkeys *(Saimiri sciureus)* as Revealed by Sociometry, Bioacoustics, and Brain Stimulation. In *Social Communication among Primates,* ed. S. A. Altmann, 149–84. Chicago: University of Chicago Press.

Ploog, D. W., and P. D. MacLean. 1963. Display of Penile Erection in Squirrel Monkeys *(Saimiri sciureus). Animal Behaviour* 11:32–39.

Poulin, R., and W. L. Vickery. 1995. Cleaning Symbiosis as an Evolutionary Game: To Cheat or Not to Cheat? *Journal of Theoretical Biology* 175:63–70.

Poulsen, B. O. 1996. Species Composition, Fluctuation and Home Range of

Mixed-Species Bird Flocks in a Primary Cloud Forest in Ecuador. *Bulletin British Ornithological Club* 116:67–74.

Prosser, C. L., ed. 1981. *Neural and Integrative Animal Physiology.* New York: John Wiley Press.

Prothero, D. R., and R. M. Schoch, eds. 1989. *The Evolution of the Perissodactyla.* New York and Oxford: Oxford University Press.

Pusey, A. E., and C. Packer. 1994. Infanticide in Lions: Consequences and Counterstrategies. In *Infanticide and Parental Care,* ed. S. Parmigiani and F. van Saal, 277–99. Switzerland: Harwood Academic Publishers.

Queller, D. C. 1995. The Spandrels of St. Marx and the Panglossan Paradox: A Critique of a Rhetorical Program. *Quarterly Review of Biology* 70:485–89.

Ralls, K. 1971. Mammalian Scent Marking. *Science* 171: 443–49.

Rand, A. S. 1988. An Overview of Anuran Acoustic Communication. In *The Evolution of the Amphibian Auditory System,* ed. B. Frish. New York: John Wiley and Sons.

Rand, A. S., and M. J. Ryan. 1981. The Adaptive Significance of a Complex Vocal Repertoire in a Neotropical Frog. *Zeitschrift für Tierpsychologie* 57:209–14.

Rasa, O. A. E. 1973. Marking Behavior and Its Significance in the African Dwarf Mongoose, *Helagale undulata rufula. Zeitschrift für Tierpsychologie* 32:449–88.

———. 1977. The Ethology and Sociology of the Dwarf Mongoose *(Helagale undulata rufula). Zeitschrift für Tierpsychologie* 43:337–406.

———. 1983. Dwarf Mongoose and Hornbill Mutualism in the Tarn Desert, Kenya. *Behavioral Ecology and Sociobiology* 12:181–90.

———. 1985. *Mongoose Watch.* London: John Murray.

———. 1994. Altruistic Infant Care or Infanticide. In *Infanticide and Parental Care,* ed. S. Parmigiani and F. S. vom Saal, 301–20. Switzerland: Harwood Academic Publishers.

Raup, D. M. 1988. Extinction in the Geologic Past. In *Origins and Extinctions,* ed. D. E. Osterbruck and P. H. Raven, 109–19. New Haven and London: Yale University Press.

Redondo, T., M. Gomendio, and R. Medina. 1992. Sex-Biased Parent-Offspring Conflict. *Behaviour* 123:261–89.

Reeve, H. K. 1992. Queen Activation of Lazy Workers in Colonies of the Eusocial Naked Mole-Rats. *Nature* 358:147–49.

Reilly, P. 1994. *Penguins of the World.* Oxford: Oxford University Press Australia.

Remsen, J. V. 1990. Community Ecology of Neotropical Kingfishers. *University of California Publications in Zoology* 134:1–116.

Reyer, H.-U. 1980a. Sexual Dimorphism and Communal Breeding in the Striped Kingfisher. *Ostrich* 51:117–18.

———. 1980b. Flexible Helper Structure as an Ecological Adaptation in the Pied Kingfisher *(Ceryle rudis rudis L.). Behavioral Ecology and Sociobiology* 6:219–27.

———. 1984. Investment and Relatedness: A Cost/Benefit Analysis of Breeding and Helping in the Pied Kingfisher *(Ceryle rudis)*. *Animal Behaviour* 32:1163–78.

Richard, A. 1978. *Behavioral Variation: Case Study of a Malagasy Lemur.* Lewisburg: Bucknell University Press.

Richardson, P. R. E. 1991. Territorial Significance of Scent-Marking during the Non-Mating Season in the Aardwolf *Proteles cristatus* (Carnivora: Protelidae). *Ethology* 87:9–27.

Richner, H., and P. Heeb. 1996. Communal Life: Honest Signaling and the Recruitment Center Hypothesis. *Behavioral Ecology* 7:115–19.

Ridley, M. 1983. *The Explanation of Organic Diversity: The Comparative Method and Adaptations for Mating.* Oxford: Clarendon Press.

Robinson, J. G. 1977. Vocal Regulation of Spacing in the Titi Monkey *Callicebus moloch*. Ph.D. diss., University of North Carolina, Chapel Hill.

———. 1979. An Analysis of the Organization of Vocal Communication in the Titi Monkey *Callicebus moloch*. *Zeitschrift für Tierpsychologie* 49 (4): 381–404.

Robinson, M. H. 1969. Defenses against Visually Hunting Predators. *Evolutionary Biology* 3:225–29.

Rodaniche, A. 1985. Notes on the Behavior of the Larger Pacific Striped Octopus *(Octopus spilotus)* Voss. *Bulletin of Marine Science* 36.

Rohwer, S. A. 1975. The Social Significance of Avian Plumage Variability. *Evolution* 29:593–610.

Roughgarden, J. 1983. Competition and Theory in Community Ecology. *American Naturalist* 122:586–601.

Rowell, T. E. 1971. Organization of Caged Groups in Cercopithecus Monkeys. *Animal Behaviour* 19:625–45.

———. 1972. *Social Behaviour of Monkeys.* London: Harmondsworth Penguin.

———. 1988. The Social System of Guenons, Compared with Baboons, Macaques and Mangabeys. In *A Primate Radiation: Evolutionary Biology of the African Guenons,* ed. A. Gauthier-Hion, F. Bourlière, J.-P. Gauthier, and J. Kingdon, 439–51. Cambridge: Cambridge University Press.

Rubenstein, D. I. 1986. Ecology and Sociality in Horses and Zebras. In *Ecological Aspects of Social Evolution: Birds and Mammals,* ed. D. I . Rubenstein and P. W. Wrangham, 282–302. Princeton: Princeton University Press.

Rubenstein, D. I., and R. W. Wrangham, eds. 1986. *Ecological Aspects of Social Evolution: Birds and Mammals.* Princeton: Princeton University Press.

Ryan, M. J. 1985. *The Tungara Frog.* Chicago: University of Chicago Press.

de Saussure, F. 1986. *Cours de linguistique générale: Edition critique préparée par Tullio de Mauro.* Paris: Payot.

van Schaik, C. P. 1989. The Ecology of Social Relationships amongst Female Primates. In *Comparative Socioecology: The Behavioral Ecology of Humans and Other Mammals,* ed. V. S. Stanton and R. A. Foley, 195–218. Oxford: Blackwell Scientific Press.

Schaller, G .B. 1963. *The Mountain Gorilla: Ecology and Behavior.* Chicago: University of Chicago Press.

———. 1972. *The Serengeti Lion.* Chicago: University of Chicago Press.

———. 1977. *Mountain Monarchs.* Chicago and London: University of Chicago Press.

Schenkel, R., and L. Schenkel-Hulliger. 1969. *Ethology and Behaviour of the Black Rhinoceros (Diceros bicornis L.).* Hamburg and Berlin: Paul Parey.

Schilder, M. B. H. 1990. Interventions in a Herd of Semi-Captive Plains Zebras. *Behaviour* 112:51–83.

Schoener, T. W. 1982. The Controversy over Interspecific Competition. *American Scientist* 70:586–95.

Schultze-Westrum, T. 1965. Innerartliche Verständigung durch Dufte beim Gleitbeutler *Petaurus breviceps* papuanus Thomas (Marsupialia, Phalangeridae). *Zeitschrift für Vergleichende Physiologie* 50:151–220.

Scott, J. P. 1958. *Aggression.* Chicago: University of Chicago Press.

Scott, P. E., and R. F. Martin. 1983. Reproduction of the Turquoise-Browed Motmot at Archaeological Ruins in Yucatan. *Biotropica* 15:8–14.

Sebeok, T. A. 1969. Semiotics and Ethology. In *How Animals Communicate,* ed. T. A. Sebeok and A. Ramsay. Bloomington: Indiana University Press.

———. 1976. *Contribution to the Doctrine of Signs.* Bloomington: Indiana University Press.

Senar, J. C., M. Camerino, and N. B. Metcalfe. 1990. Familiarity Breeds Tolerance: The Development of Social Stability in Flocking Siskins *(Carduelis spinus). Ethology* 85:13–24.

Seyfarth, R. 1976. Social Relations among Adult Female Baboons. *Animal Behaviour* 24:917–38.

Sherman, P. J. 1995. Breeding Biology of White-Winged Trumpeters *(Psophia leucoptera)* in Peru. *Auk* 112:285–95.

Sherman, P. W. 1977. Nepotism and the Evolution of Alarm Calls. *Science* 197:1246–53.

Sherman, P. W., J. V. M. Jarvis, and R. D. Alexander, eds. 1991. *The Biology of the Naked Mole Rat.* Princeton: Princeton University Press.

Sherman, P. W., J. V. M. Jarvis, and S. H. Braude. 1992. Naked Mole Rats. *Scientific American* 267 (2): 72–78.

Short, L. L. 1973. Habits of Some Asian Woodpeckers (Aves, Picidae). *Bulletin of the American Museum of Natural History* 152:253–364.

———. 1979. Burdens of the Picid Hole-Excavating Habit. *Wilson Bulletin* 91:16–28.

———. 1982. *Woodpeckers of the World.* Delaware Museum of Natural History Monograph 4. Wilmington.

Shuttleworth, S. 1995. Mechanistic Behavioral Ecology. *Trends in Ecology and Evolution* 10:177–78.

Sibley, C. G., and J. E. Ahlquist. 1972. *A Comparative Study of the Egg White Proteins of Non-Passerine Birds.* Peabody Museum of Natural History Yale University Bulletin 39. New Haven.

———. 1985. The Relationships of Some Groups of African Birds, Based on Comparisons of the Genetic Material, DNA. In *Proceedings, International Symposium, African Vertebrates, Bonn,* 115–61.

———. 1990. *Phylogeny and Classification of Birds.* New Haven and London: Yale University Press.

Sigogneau-Russell, D. 1991. *Les Mammifères au Temps des Dinosaures.* Paris: Masson.

Silk, J. B. 1994. Social Relationships of Male Bonnet Macaques: Male Bonding in a Matrilineal Society. *Behaviour* 130:271–91.

Silverberg, J., and J. P. Gray, eds. 1992. *Aggression and Peacefulness in Humans and Other Primates.* New York: Oxford University Press.

Skutch, A. F. 1954. *Life Histories of Central American Birds: Pacific Coast Avifauna.* Cooper Ornithological Society 31. Berkeley.

———. 1957. Life History of the Amazon Kingfisher. *Condor* 59:217–29.

———. 1967. *Life Histories of Central American Highland Birds.* Publications of the Nuttall Ornithological Club 7. Cambridge, England.

———. 1983. *Birds of Tropical America.* Austin: University of Texas Press.

Slobodchicoff, C. N., ed. 1988. *The Ecology of Social Behavior.* San Diego: Academic Press.

Smith, J. L., and H.-T. Yu. 1992. The Association between Vocal Characteristics and Habitat Type in Taiwanese Passerines. *Zoological Science* 9:659–64.

Smith, W. J. 1965. Message, Meaning, and Context in Ethology. *American Naturalist* 99:405–9.

———. 1977. *The Behavior of Communicating.* Cambridge: Harvard University Press.

———. 1985. Consistency and Change in Communication. In *The Development of Expressive Behavior, Biology-Environment Interactions,* ed. G. Ziven. New York: Academic Press.

Smith, W. J., S. L. Smith, J. G. de Villa, and E .L. Oppenheimer. 1976. The Jump-Yip Display of the Black-Tailed Prairie Dog, *Cynomys ludovicianus. Animal Behaviour* 24:609–21.

Smith, W. J., S. L. Smith, E. L. Oppenheimer, and J. G. de Villa. 1977. Vocalizations of the Black-Tailed Prairie Dog, *Cynomys ludovicianus. Animal Behaviour* 25:152–64.

Smuts, B. B. 1985. *Sex and Friendship in Olive Baboons.* New York: Aldine.

Smythe, N. 1977. The Function of Mammalian Alarm Advertising: Social Signals or Pursuit Invitation? *American Naturalist* 111:191–94.

Sowls, L. K. 1984. *The Peccaries.* Tucson: University of Arizona Press.

Spinage, C. 1994. *Elephants.* London: Academic Press.

Spurway, H., and J. B. S. Haldane. 1953. The Comparative Ethology of Vertebrate Breathing. *Behaviour* 6:1–34.

Stacey, P. B., and W. D. Koenig, eds. 1990. *Cooperative Breeding in Birds.* Cambridge: Cambridge University Press.

Starin, E. D. 1994. Philopatry and Affiliation among Red Colobus. *Behaviour* 130:253–70.

St. Clair, C. C., J. R. Waas, R. C. St. Clair, and R. T. Boag. 1995. Unfit Mothers? Maternal Infanticide in Royal Penguins. *Animal Behaviour* 50:1177–85.

Stokes, A. W. 1962a. Acoustic Behaviour among Blue Tits at a Winter Feeding Station. *Behaviour* 19:118–38.

———. 1962b. The Comparative Ethology of Great, Blue, Marsh and Coal Tits at a Winter Feeding Station. *Behaviour* 19:208–18.

Straffin, P. D. 1993. *Game Theory and Strategy.* Mathematical Association of America, New Mathematical Library.

Strier, K. B. 1994. Brotherhoods among Atelins: Kinship, Affiliations and Competition. *Behaviour* 130:151–67.

Susman, R. L. 1984. *The Pygmy Chimpanzee: Evolutionary Biology and Behavior.* New York and London: Plenum Press.

Svensson, E. 1995. Parent-Offspring Relations in Mammals. *Trends in Ecology and Evolution* 10:83.

Tanaka, Y. 1996. Social Selection and the Evolution of Animal Signals. *Evolution* 50:512–23.

Tardif, S. D. 1994. Relative Energetic Cost of Infant Care in Small-Bodied Neotropical Primates and Its Relation to Infant-Care Patterns. *American Journal of Primatology* 34 (2): 133–43.

Tattersall, I. 1982. *The Primates of Madagascar.* New York: Columbia University Press.

Tembrock, G. 1974. Sound Production of *Hylobates* and *Symphalangus.* In *Gibbon and Siamang,* vol. 3. Basel: Karger.

Tenaza, P. R. 1976. Songs, Choruses and Countersinging of Kloss' Gibbon *(Hylobates klossii)* in Siberut Island, Indonesia. *Zeitschrift für Tierpsychologie* 40 (1): 37–52.

van Tets, G. F. 1965. A Comparative Study of Some Social Communication Patterns in the Pelecaniformes. Ornithological Monographs 2:1–88. Washington, D.C.: American Ornithological Union.

Thiollay, J. M. 1985. Stratégies adaptives comparées des rolliers sédentaires et migrateurs dans une savanne guinéene. *Terre et la Vie* 40:355–78.

Thorpe, W. H. 1972. The Comparison of Vocal Communication in Animals and Man. In *Non-Verbal Communication,* ed. R. A. Hinde, 27–47. Cambridge: Cambridge University Press.

Thresher, R. E. 1984. *Reproduction in Reef Fishes.* Neptune City, N.J.: T.F.H. Publications.

Tiger, L. 1993. Biological Antecedents of Human Aggression. In *Primate Social Conflict,* ed. W. A. Mason and S. P. Mendoza, 373–86. Albany: State University of New York Press.

Tinbergen, N. 1940. Die Übersprungbewegung. *Zeitschrift für Tierpsychologie* 4:1–40.

———. 1951. *The Study of Instinct.* London: Clarendon Press.

———. 1952a. *The Herring Gull's World: A Study of the Social Behaviour of Birds.* London: Collins.

———. 1952b. "Derived" Activities: Their Causation, Biological Significance, Origin, and Emancipation during Evolution. *Quarterly Review of Biology* 27:1–32.

———. 1960. Comparative Studies of the Behaviour of Gulls (Laridae): A Progress Report. *Behaviour* 15:1–70.

———. 1963. On Aims and Methods of Ethology. *Zeitschrift für Tierpsychologie* 20:410–33.

Toonan, R. J., and J. R. Pawlik. 1994. Foundations of Gregariousness. *Nature* 370 (6490): 511–12.

Trivers, R. 1974. Parent-Offspring Conflict. *American Zoologist* 14:249–65.

———. 1985. *Social Evolution.* Menlo Park, Calif.: Benjamin Cummings.

Turner, L. C. F. 1985. The Significance of the Schlieffen Plan. In *The War Plans of the Great Powers, 1880-1994,* ed. P. M. Kennedy, 199–221. Boston: Allen and Unwin.

Urban, E. K., C. H. Fry, and S. Keith, eds. 1986. *The Birds of Africa.* Vol. 2. New York: Academic Press.

Vasey, P. L. 1995. Homosexual Behavior in Primates: A Review of Evidence and Theory. *International Journal of Primatology* 16:173–204.

Vehrencamp, S. L. 1978. The Adaptive Significance of Communal Nesting in Groove-Billed Anis *(Crotophaga sulcirostris). Behavioral Ecology and Sociobiology* 4:1–33.

———. 1979. The Roles of Individuals, Kin and Group Selection in the Evolution of Sociality. In *Handbook of Behavioral Neurobiology,* vol. 3, ed. P. Marler and J. G. Vandenbergh, 351–94. New York: Plenum Press.

Vermeigh, G. J. 1987. *Evolution and Escalation: An Ecological History of Life.* Princeton: Princeton University Press.

Veuilleumier, F., and M. Monasterio, eds. 1986. *High Altitude Tropical Biogeography.* Oxford: Oxford University Press.

Vogel, C. 1976. *Ökologie, Lebensweise und Sozialverhalten der Grauen Languren in Verschiedenen Biotopen Indiens.* Advances in Ethology 17. Berlin and Hamburg: Paul Parey.

de Waal, F. B. M. 1986. Deception in the Natural Communication of Chimpanzees. In *Deception: Perspectives on Human and Nonhuman Deceit,* ed. R. W. Michell and Nick S. Thompson, 221–44. Albany: State University of New York Press.

———. 1988. The Communicative Repertoire of Captive Bonobos *(Pan paniscus)* Compared to that of Chimpanzees. *Behaviour* 106:183–251.

———. 1989. *Peacemaking among Primates.* Cambridge: Harvard University Press.

———. 1995. Bonobo Sex and Society. *Scientific American* (March): 82–83.

de Waal, F. B. M., and D. Yoshihara. 1985. Reconciliation and Re-Directed Affection in Rhesus Monkeys. *Behaviour* 83:224–41.

Wallace, M. P., and S. A. Temple. 1987. Competitive Interactions within and between Species in a Guild of Avian Scavengers. *Auk* 104:290–95.

Walther, F. R. 1984. *Communication and Expression in Hoofed Mammals.* Bloomington: Indiana University Press.

Watts, C. R., and A. W. Stokes. 1971. The Social Order of Turkeys. *Scientific American* 224:112–18.

Watts, D. P. 1995a. Post-Conflict Social Events in Wild Mountain Gorillas (Mammalia, Hominoidea). 1. Social Interactions between Opponents. *Ethology* 100:139–57.

———. 1995b. Post-Conflict Social Events in Wild Mountain Gorillas. 2. Redirection, Side Direction, and Consolation. *Ethology* 100:158–74.

Wells, K. D. 1977. The Social Behavior of Anuran Amphibians. *Animal Behaviour* 25:666–93.

Wells, M. J. 1978. *Octopus: Physiology and Behaviour of an Advanced Invertebrate.* London: Chapman and Hall.

Whitman, C. O. 1919. The Behavior of Pigeons. *Publications of the Carnegie Institution* 257:1–161.

Whittingham, L. A., P. O. Dunn, and R. J. Robertson. 1995. Testing the Female Mate-Guardian Hypotheses: A Reply. *Animal Behaviour* 50:277–79.

Wickler, W. 1963. Die Biologische Bedeutung Auffallend Farbiger Nackter Hautstellen und Innerartliche Mimicry der Primaten. *Naturwissenschaften* 50:481–82.

———. 1967. Sociosexual Signals and Their Interspecific Imitation among Primates. In *Primate Ethology,* ed. D. Morris, 69–147. London: Weidenfeld and Nicolson.

———. 1968. *Mimicry in Plants and Animals.* New York: McGraw-Hill.

Williams, G. C. 1966. *Adaptation and Natural Selection: A Critique of Some Current Evolutionary Thought.* Princeton: Princeton University Press.

Williams, H., and F. Nottebohm. 1985. Auditory Responses in Avian Vocal Motor Neurons: A Motor Theory for Song Perception in Birds. *Science* 229:279–82.

Williams, J. G. 1964. *A Field Guide to the Birds of East and Central Africa.* Boston: Houghton Mifflin.

Wilson, E. O. 1975. *Sociobiology: The New Synthesis.* Cambridge: Belknap Press of Harvard University Press.

Wilson, R. T. 1992. Morphology and Physiology of the Hammerkop *Scopus umbretta* Egg. In *Proceedings, 7th Pan-African Ornithological Congress,* 337–42. Nairobi.

Wolf, L. L., and F. R. Hainsworth. 1971. Time and Energy Budgets of Territorial Hummingbirds. *Journal of Ecology* 52:980–88.

Wolf, L. L., F. R. Hainsworth, and F. G. Stiles. 1972. Energetics of Foraging Rate and Efficiency of Nectar Extraction by Hummingbirds. *Science* 176:1351–52.

Woolfenden, G. E., and J. W. Fitzpatrick. 1984. *The Florida Scrub Jay.* Princeton: Princeton University Press.

Wrege, P. H., and S. T. Emlen. 1994. Family Structure Influences Mate Choice in White-Fronted Bee-Eaters. *Behavioral Ecology and Sociobiology* 35:185–91.

York, A. D., and T. E. Rowell. 1988. Reconciliation Following Aggression in Patas Monkeys, *Erythrocebus patas. Animal Behaviour* 36:502–9.

Young, J. Z. 1977. Brain, Behaviour and Evolution of Cephalopods. In *The Biology of Cephalopods,* ed. M. Nixon and J. B. Messenger. London: Academic Press.

Young, R. E. 1981. Color of Bioluminescence in Pelagic Organ Tissue. In *Bioluminescence: Current perspectives,* ed. K. Malson. Minneapolis: Burgess.

Zahavi, A. 1975. Mate Selection: A Selection for a Handicap. *Journal of Theoretical Biology* 53:205–14.

———. 1977. The Cost of Honesty (Further Remarks on the Handicap Principle). *Journal of Theoretical Biology* 67:603–5.

———. 1987. The Theory of Signal Selection. In *Proceedings of the International Symposium on Biological Evolution,* ed. V. P. Delfino, 305–25. Bari: Adriatica Edetricia.

———. 1991. On the Definition of Sexual Selection, Fisher's Model, and the Evolution of Waste and of Signals in General. *Animal Behaviour* 42:502–3.

Zahn-Waxler, C., E. M. Cummings, and R. Iannoti. 1986. Introduction. In *Altruism and Aggression: Biological and Social Origins,* ed. C. Zahn-Waxler, E. M. Cummings, and R. Ianotti. Cambridge: Cambridge University Press.

Zink, R. M. 1996. Bird Species Diversity. *Nature* 381 (6583): 566.

INDEX

Page numbers followed by *f* refer to figures.